ISLANDS IN DEEP TIME

OTHER BOOKS

By Markes E. Johnson

Discovering the Geology of Baja California: Six Hikes on the Southern Gulf Coast (2002)

Off-Trail Adventures in Baja California: Exploring Landscapes and Geology of Gulf Shores and Islands (2014)

Baja California's Coastal Landscapes Revealed: Excursions in Geologic Time and Climate Change (2021)

ISLANDS IN DEEP TIME

Ancient

Landscapes

Lost and Found

MARKES E. JOHNSON

Columbia University Press

New York

Columbia University Press
Publishers Since 1893
New York   Chichester, West Sussex
cup.columbia.edu

Copyright © 2023 Columbia University Press

Cataloging-in-Publication data are available from the Library of Congress.
ISBN 978-0-231-21218-2 (hardback)
ISBN 978-0-231-21219-9 (trade paperback)
ISBN 978-0-231-55925-6 (ebook)

Printed in the United States of America

Cover design: Philip Pascuzzo
Cover images: Map, elements from T. C. Chamberlin (1881), *Hypothetical Map of the Currents of the Silurian Interior Sea*. Photo: Shuttersock.

DEDICATED TO THE MEMORY of Rachel Carson (1907–1964), whose descriptions of shorelines and ocean processes are unsurpassed in elegance. Her outlook was on our present world. But in her writings, she pointed out at nearly every turn of the page that the world's shores and oceans have a geologic past stretching far back in time.

CONTENTS

....

PREFACE

On the Reality of Time Travel

Compared to entire continents, islands are miniature worlds with discrete ecologies that are both more vulnerable to environmental disruption and more fertile as incubators of biological innovation. This book offers excursions to a dozen ancient islands caught in geologic time with relationships exceptionally preserved among fossil plants and animals in the context of the coastal settings they once inhabited. Each chapter includes a topographic map representing the island's physical structure paired with a global map showing its location in the context of the geography unique to that time. Through succeeding chapters, these maps trace the changing positions of continents and ocean basins beginning five hundred million years ago to the present day. How a given island at a specific spot was buffeted by prevailing winds, waves, and storms that sculpted rocky shores affected the range of marine invertebrates and certain algae best suited to thrive under those conditions. The accompanying narrative provides a guided tour around each island—or, in some cases, over the top of that island—augmented along the way by images detailing fossil finds. The result is a window into the evolving complexity of coastal ecology through time and the ways life spread around the world aided by island stepping stones across the seas of the blue planet we call Earth.

The only qualification for anyone picking up this book is the curiosity and sense of adventure that attract visitors to U.S. national parks, especially those organized around islands. They might include Mount Dessert Island in Acadia National Park of Maine, Saint John Island in the Virgin Islands National Park, the Channel Islands National Park off the coast of southern California, and Volcanoes National Park on the Big Island of Hawaii. All are mentioned in this book, but key chapters focus more intently on smaller parks in many parts of the world. In particular, the guiding lodestone for the journeys advocated herein is Mount Monadnock in southern New Hampshire, which lends its geographic name (*monadnock*) to structures worldwide, both past and present, that commonly alternate between watery islands and sky islands left high and dry by tectonic forces or the sea's periodic retreat. Those having some familiarity with such places will enlarge their understanding of the duality of landscapes that juxtapose the past and the present at the same spot, often in startling ways.

Ecologic time and geologic time are different. When climbing a mountain like Cadillac Mountain in Acadia National Park or making landfall on an island like Santa Cruz in the Channel Islands National Park, we participate in ecologic time. There is the moment-by-moment lapse of time from one sunrise to the following sunset and on again, during which diurnal tides rise and fall and all creatures go about the normal business of living, diurnal, nocturnal, or a mix of the two. The crux of this book is bound to the idea that we may inhabit two different worlds at the same time. Anywhere we might go, the local landscape is the stage on which ecological relationships among living plants and animals take place. We may explore a contemporary landscape on a perfectly sunny day governed by ecologic time and simultaneously enter a bygone world faithfully preserved as a rare slice of ecologic time from the distant past. That is to say, we humans possess an ability by which we may journey to a place with one eye on the present and the other trained to recognize and enter a former world preserved in intricate detail beneath the skin of the same landscape. Time travel is a reality in this context routinely subscribed to by geologists.

Fundamental questions come to mind regarding the natural history of islands. What makes one island different from another? What differences do the foundational rocks of any particular island make to the life expectancy of that island? How does life first arrive on fresh island shores? What factors allow some marine species, such as corals, oysters, or barnacles, to cross deep oceans, hopping from one island to another largely unchanged, but stopping others short? What circumstances impel some colonists to evolve into new species once the first foothold is gained? Why do islands disappear? Geologic time is often referred to as deep time because of the immense age of planet Earth, and that context makes a great difference in how such questions are addressed. Fossil islands, as they may be considered, promise to yield answers to such questions that cannot easily be answered by present-day islands. Continental shelves and parts of the ocean floor are subject to uplift and drainage that otherwise achieves the same end as a dry dock. When a ship is brought into a dry dock, it is left high and dry so that the entire structure is exposed for examination and required repairs. In short, dry-docking allows for the examination of paleoislands below their former waterline in ways otherwise impossible.

Planet Earth is old, as much as 4.5 billion years. Its islands have a history readily recognized as much as five hundred million years ago, during the Cambrian period (chapter 2). Examples of actual islands buried and preserved in the rock record can be identified because they were exhumed later by natural erosion and became available for exploration in the style of time travel habitual to geologists. Astonishing as it may seem, Earth's watery surface has waxed and waned over the eons. At times during the Cambrian period, only 10 or 15 percent of North American land surface was exposed. At other times, 250 million years ago, when sea level was at a low ebb and all continents were assembled as a single entity that we call Pangea, the land and its surrounding outer shelf may have totaled 40 percent of the planet's surface. Today, water covers 70 percent of Earth's surface. Whether affixed to continents as stable monadnocks or rooted in the crust of shifting oceans as volcanic peaks, island life persisted and diversified as it swept from place to place.

Books matter, especially in their power to influence young and old alike. As a teenager growing up during the height of the Cold War, I was surprised to find a book in my local public library translated from Russian, authored by a Moscow State University professor named R. F. Hecker. This slim book, *Introduction to Paleoecology* (1965), explained how to see fossils in another light, as part of former ecosystems. Hecker explored places like the Jurassic cliffs along the lower Volga River in southern Russia, the Jurassic "paper shales" of Kazakhstan, and the much younger Paleogene rocks of Turkestan. Most intriguingly, he described those places as bearing fossils in such a perfect state of ecological detail that they amounted to outdoor museums worthy of celebration as natural monuments. Hecker's foresight anticipated the global geopark movement launched in 2001 by the United Nations Educational, Scientific, and Cultural Organization (UNESCO) to foster outdoor museums in promoting our common geoheritage.

In seeking paleoislands across North America and elsewhere around the world, my aim has been to identify places worthy of induction as natural monuments. Some I stumbled onto as a result of pure, dumb luck. Chapter 3 describes an archipelago that dates back some 445 million years to the close of the Ordovician period. It is located near the town of Churchill, Manitoba, on the shores of Canada's Hudson Bay. I went overland by rail to Churchill in 1984 on a lark after finishing another assignment in central Manitoba. On reaching the shores of the great bay, with small icebergs still afloat at the end of July, I let out a whoop of excitement that startled my student assistants. "What is so special?" they wanted to know. By serendipity, we had arrived at the coast during low tide. The modern coastline could be seen as congruent with an ancient rocky shore represented by enormous boulders of dark quartzite embedded in sedimentary layers of beige dolostone. At high tide, the scene vanished beneath the waves leaving only the

parent monadnock in sight. I tried to explain how unusual I thought the relationship was. I was instantly challenged by my students to prove that other such
places were perhaps more common than I thought.

The effort in finding and getting to those other places, like the Devonian
monadnock islands on the flanks of the Kimberley Range in Western Australia
from 375 million years ago (chapter 5), was deliberate and planned in great detail.
On the ground, it required days of walking to put together a lucid story. The tales
told by paleoislands that survive as potential natural monuments often record
big events like major storms and even hurricanes. Were hurricanes more frequent
and more powerful in the past than today? Perhaps not always so, but during the
Pliocene warm period described in relation to paleoislands of that age between
five and three million years ago (chapter 10), they almost certainly were. Indeed,
the Pliocene warm period has been held up as a cautionary warning about the
future of our planet under threat of rapid and pervasive global warming. The
geological time chart is cumbersome, and it takes some getting used to. But destinations in time, like the Cambrian, Ordovician, Devonian, or Pliocene, follow
conventional signposts established by geologists to plot our way to other worlds
where we may experience moments of ecologic time that rival those in familiar
landscapes today.

Geologists and paleontologists have a vital role to play by showing how the
natural mood swings of our planet occurred long before humans were around.
If we can better understand the physical conditions that contributed to those
swings, we can at least make a better prediction as to what lies ahead due to
global warming. Beyond advocacy for a cleaner planet, the goal of this book is
to promote ways we may enjoy nature through geotourism in celebration of our
geoheritage through the sustainable economic development of geoparks and
palaeoparks. An ancillary question about what uses we make of islands and how
we come and go from islands is addressed in the closing remarks (chapter 12),
emphasizing monadnocks as places for something akin to spiritual reflection.

With that epilogue in mind, I invite you to join me on journeys through deep
time across the geography of today's world to land on paleoislands, all of which
satisfy the high standards of natural monuments. Mount Monadnock and its
flooded twin on Hongdo (Red Island) in the Yellow Sea off the Korean peninsula (chapter 1) are our training grounds for a prodigious adventure in time and
planetary introspection. What awaits is a sampling of ancient islands secured in
dry dock where they may be explored at ease above and below their former water
lines. We will visit them in a sequence that follows through time from a seascape
a half-billion years ago to our own doorstep. Science has provided us with an
amazing set of tools, from satellite reconnaissance at one extreme to the electron

microscope at the other, both of which I have used in my studies. But the kind of adventure I advocate here can be done with the simplest tools readily available to all: a rock hammer, hand lens, and topographic map. My hope is you will begin to think about islands differently. Perhaps we may grow to better appreciate our planetary home fixed in the vast universe as a small and vulnerable island worthy of greater reverence and a renewed dedication to stewardship.

ACKNOWLEDGMENTS

Foremost, Williams College provided a stable environment for teaching and research over a thirty-five-year career that extended into retirement, with additional support through the Office of the Dean of Faculty. A generous subsidy toward publication of this book came from the Class of 1958 Fund for Faculty Enhancement at Williams College. Among a score of talented students, Hal Lescinsky, Ken MacLeod, and David Skinner accompanied the author to the shores of Hudson Bay at Churchill, Manitoba, to investigate Ordovician and Silurian paleoislands. On separate occasions, Dan Walsh and Jim Scott traveled to Western Australia to take part in studies on Devonian and Permian paleoislands, respectively. Marshall Hayes, Hovey Clark, and Jon Payne came to Mexico's Baja California peninsula to study the Cretaceous paleoislands around Eréndira on the Pacific Coast. Max Simian, Mike Eros, and Dan Perez made multiple trips to the Gulf of California to map Pliocene paleoislands, and Laura (Libby) Blackmore, Cordelia Ransom, Patrick Russell, and Peter Tierney did likewise with respect to Pleistocene paleoshores. In addition to their welcome company, these outstanding Williams students made independent contributions through their own vigilant field observations.

Support for field studies in China's Inner Mongolia and in Western Australia came from the National Geographic Society through separate grants that enabled the material covered in chapters 4, 5, and 6 to be developed in full. Several grants from the Petroleum Research Fund through the American Chemical Society supported fieldwork on Cretaceous and Pliocene paleoislands covered in chapters 8 and 10. A grant from the Marion and Jasper Whiting Foundation enabled travel to South Korea's Hongdo (Red Island), described in chapter 1. Three years of grant support through the Spanish Ministry of Science and

Innovation to Eduardo Mayoral (Universidad de Huelva) supported field studies in the Madeira and Cape Verde Islands, described in chapters 9 and 11. As related in chapter 10, I was fortunate to participate in multiple summer sessions studying Pliocene and Pleistocene paleoshores on Santa Maria Island in the Azores under generous town- and provincial-level support for the international workshop Paleontology in Atlantic Islands, orchestrated by Sérgio Ávila (Universidade dos Açores).

Research colleagues from other institutions combined forces with me to carry out projects far afield. A long-time field partner, Jorge Ledesma-Vázquez (Universidad Autónoma de Baja California) hosted projects in Mexico. Rong Jia-yu (Nanjing Institute of Geology and Palaeontology) sponsored multiple visits to China and participated in fieldwork on Ontario's Manitoulin Island as well as the shores of Manitoba's Hudson Bay. Gregory Webb (University of Queensland) was the indispensable coordinator of field studies in the Kimberley district of Western Australia. The late Stuart W. McKerrow (Oxford University) encouraged and engaged with fieldwork on the Jurassic of south Wales and Somerset. Both from a scientific and cultural point of view, little could surpass the fellowship enjoyed with colleagues Carlos Marques da Silva and Mário Cachão (Universidade de Lisboa), Ana Santos and Eduardo Mayoral (Universidade de Huelva), and Ricardo Ramalho (Cardiff University) during adventures in the Macaronesian Islands of the North Atlantic. Kenneth Donovan kindly scoured the manuscript for technical errors and oversights that were difficult for the author to spot without assistance. Thanks are owed to Paul D. Taylor, editor of the journal *Palaeontology*, for permission to reprint figures in chapter 7 originally published there.

My partner in all things domestic and foreign, B. Gudveig Baarli, deserves the highest praise as a scientific collaborator and for her ingenuity as a travel agent, technical adviser on computer draftsmanship, and steadfast life companion. May we find many more islands to enjoy together, including those on your unfinished wish list.

ISLANDS IN DEEP TIME

CHAPTER 1

HOW TO LISTEN TO A SKY ISLAND
WITH GLOBAL AMBITION

Climbing Mount Monadnock

On the summit as I stood,
O'er the wide floor of plain and flood,
Seemed to me the towering hill
Was not altogether still,
But a quiet sense conveyed.

—Ralph Waldo Emerson, "Monadnoc" (1845)

There exists about the place a many-sided and timeless riddle. Untold numbers of repeat visitors come during all seasons of the year. They explore the landscape's craggy face to stretch muscles in the gladness of being outdoors in nature and to exercise the mind in wonderment at such a place. The riddle is in three parts: What rises head and shoulders above its neighbors but rests on folded limbs at an unfathomable depth? What is mute but wears a cloak with a written language of its own? What constitutes a singularity unto itself yet exudes a simultaneous presence over other parts around the world? The answer is Mount Monadnock. It is a single peak in southern New Hampshire, the summit of which registers a modest elevation at 3,165 feet (965 m) above sea level. The name is said to derive from a native Abenaki phrase for "mountain that stands alone." Mount Monadnock rises 2,000 feet (610 m) higher than any other prominence within a 30-mile (48 km) radius, but its name answers only part of the riddle.

Near my home in western Massachusetts, Mount Monadnock can be spied as a tiny blip on the eastern horizon viewed from a roadside spot at the crest of the Hoosac Range 56 miles (90 km) away. Closer, at 34 miles (55 km), the spot appears more substantial, as seen from Hogback Mountain in southern Vermont. On the road's gentle descent leading to Keene, New Hampshire, the mountain takes on its dominant profile above the immediate countryside. I have lost track of how often the mountain has drawn me to its slopes. Sometimes I attend during the summer, more often during the autumn when the forest foliage is colorful, but only seldom on the edge of winter when the trails are spotted with ice before the first snowfall. The contrasting seasons cast the mountain in different moods, most telling during early summer when bands of vegetation on the slopes are refreshed with new growth. Striking as a vestment, the plant life adds to the multipart riddle. In effect, Mount Monadnock is a sky island where certain plants that thrive at higher altitudes are isolated from surrounding lowlands unfavorable to their growth.

As a geologist, I view any landscape with a set of basic questions: What kinds of rocks underlie a particular landscape? How did they get there (i.e., what processes left those rocks in place)? Finally, what is their significance in terms of a recognizable pattern that conforms to landscapes elsewhere? Such queries offer a logical formula to address the greater riddle of Mount Monadnock. The exercise yields maximum pleasure when the questioner directly interacts with the object under consideration. But we humans can use our imagination, stimulated by the printed word and other appropriate iconography. Case in point, a topographic map provides the means to compress a large feature onto a small piece of paper supplemented with descriptive notes as guideposts. Something larger than the 5,000 acres (~2,000 ha) of uplands around the mountaintop is represented by such a map (fig. 1.1). The climb up Mount Monadnock is said to rank third in popularity as a day excursion, after Japan's Mt. Fuji and China's Tai Shan. Many have favorite routes that lead to the summit, and I am no exception. My visits begin early in the day, taking the most direct but steepest route on the White Arrow trail up the south face.

FEET ON THE GROUND

An encounter with Mount Monadnock can be considered a kind of training exercise for the deeper journeys in time and space ahead. Concentric topographic lines on the map (fig. 1.1) outline the central core of Mount Monadnock, with brooks flowing outward in a radial pattern around the circumference. Modest V-shaped indentations etched by each rivulet through all lines of equal elevation show that erosion is taking place, albeit minor in impact. Our hike starts from

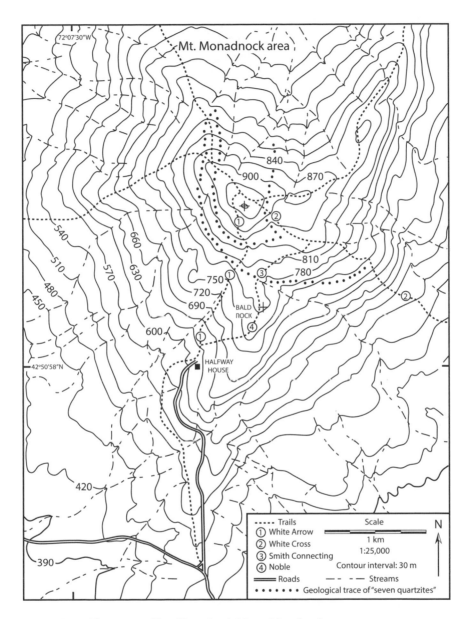

FIGURE 1.1 Elevations on New Hampshire's Mount Monadnock.

the highway at an elevation of 1,300 feet (400 m) above sea level and portends an ascent of 1,850 feet (565 m). The first mile and a half (2 km) ascend through lowland forest adjacent to the Halfway House road on a gentle incline skirting the lower southwest side of the mountain. Dominant trees include the sugar maple

(*Acer saccharum*) and beech (*Fagus grandifolia*). Exposed bedrock appears after we merge above the old toll road and meet the White Arrow trail. The original Halfway House was built as a hotel in 1858 and became an immediate success, which led to an expansion with stables to accommodate as many as one hundred guests and their horses.

Here, at an elevation of 2,035 feet (620 m), the clearing reveals solid bedrock. The rock is a metamorphic product of mud and silty sediments altered to schist under heat and pressure. Streaks appear on the polished surface as traces of minerals segregated into the thinnest of laminations (foliations to a geologist). Other lines that cross the surface askew are scratches (striations) left from the gouging ice that crossed the mountain during the last great glaciation, which ended fifteen thousand years ago. Thus, the mountain also was eroded by ice during relatively recent times. Assigned to the Littleton Formation, the original sediments that preceded metamorphic alteration accumulated under marine conditions roughly four hundred million years ago during the early part of the Devonian period.[1]

Continuing on the White Arrow trail, blocks of schist loosened at the surface exhibit a pattern some have likened to turkey tracks. To my eye, they are more like the markings on rune stones left by early Norse people as monuments in Scandinavia. Like the rune stick figures, they may also constitute writings because they detail the conditions under which metamorphism occurred. The patterns are pseudomorphs of the mineral sillimanite.[2] Burial at a depth between 6 and 12 miles (10 and 20 km) under temperatures over 900°F (500°C) led to their development. Deep burial and subjection to high pressure were caused by the Acadian orogeny. It was a time when the proto–Atlantic Ocean between ancestral North America and northern Europe vanished, and the two paleocontinents collided to form a unified Euramerica. This resulted in the Acadian phase of mountain building in the Appalachians of the eastern United States and the equivalent Caledonide phase in Scotland and Norway, formerly one continuous mountain chain.

Ahead, the White Arrow mounts a succession of rock stairs (fig. 1.2) built by the Civilian Conservation Corps during the Great Depression in the 1930s. Some larger blocks are composed of quartzite, another metamorphic product caused by the fusion of ordinary sand grains under heat and pressure to make a dense and extremely hard rock. The stairs are artificial, of course, but lend a surreal effect, as if crossing hallowed ground to a mountain temple. The surrounding forest has shifted to evergreen trees dominated by red spruce (*Picea rubens*). Between 2,560 feet (780 m) and 2,755 feet (840 m), the trail passes through one of several folds in the rocks indicated by the repetition of thin quartzite beds, only 4 to 12 inches (10 to 30 cm) in thickness. These constitute a marker for the

FIGURE 1.2 Rock stairway on the White Arrow trail built by CCC workers in the 1930s.

Photo by author.

seven quartzites traced in exacting detail across much of Mount Monadnock. Overall, they relate how the rocks of the Littleton Formation were repeatedly bent backward and forward on themselves in great folds. Under compression at great depth, the rocks behaved much like toffee candy, fully malleable in a warm state before cooling.

Long flights of steps rise steeply to cross through a succession of quartzite benches, each offering a welcome pause for rest. At about 2,820 feet (860 m), the White Arrow clears a massive shelf formed by thick quartzite beds that tilt gently northeast. Open to the sky among thinning spruce trees, the beds overlap like playing cards in a splayed deck to make space for a series of shallow pools.

The pools are small bogs during the summer, an ideal habitat at this elevation for the tussock cotton sedge (*Eriophorum vaginatum*). The encounter during my first visit came as a surprise because I was accustomed to seeing the same cotton grass along the Arctic coast of Norway. With the mountain cranberry (*Vaccinium vitis-idaea*), such plants are local holdouts in an alpine setting isolated from relatives in more northerly locals. Spruce trees are more stunted and scattered higher up, allowing for a better view of the mountaintop (fig. 1.3). Exposed quartzite layers exhibit much the same orientation as at the ledge with the cotton grass but stack cumulatively through the upper mountain slope.

FIGURE 1.3 Upper part of Mount Monadnock on the White Arrow trail with sparse cover of red spruce.

Photo by author.

This alignment led to large-scale erosion by the southward flow of ice during the last glaciation. In profile, the gentle inclination in the top layer of each ledge offered little resistance to the overriding weight of thick ice, but the vertical face of each ledge was susceptible to downward plucking action. The term for this phenomenon is a *roche moutonnée*, also called a sheepback. The quaint expression was handed down from the earliest students of mountain glaciers in Switzerland from the early 1700s. The final ascent to the summit of Mt. Monadnock is steep, crossing bare rock. An elegant slender cairn (fig. 1.4) stands near the junction of our trail with those that descend the mountain to the west and north.

FIGURE 1.4 Rock cairn on the White Arrow trail near junction of trails descending Mount Monadnock to the north and west.

Photo by author.

On May 3, 1845, the poet-philosopher Ralph Waldo Emerson (1803–1882) sat near the summit and wrote a poem of 160 lines under the title "Monadnoc." The piece was published in a collection of the author's poems in 1847 and did much to fuel the mountain's popularity as a hiking destination. By then, Emerson had already cultivated a reputation as the founding leader of transcendentalism.[3] Among the notables joining the group was Henry David Thoreau (1817–1862), who climbed Mount Monadnock on several occasions between 1844 and 1860. At the heart of transcendentalism is a belief that the spirit of divinity is reflected everywhere in all of nature and humanity. The mountain spoke to Emerson, and through him it gave voice to the notion of a physical world inviolably sacred. More than a few visitors to the mountain left behind rock carvings to record their names and the year of their ascent. The oldest such inscription is said to be from 1801. The oldest I have found is from 1816.

In midsummer, the mountaintop is a magical place to enjoy the scenery laid out below on all sides. When windy, it is prudent to take lunch in a sheltered pocket below the crest of the mountain. My preferred path for the return hike seeks a less precipitous descent on the White Cross and Smith Connecting trails that convey the hiker to Bald Rock at an elevation of 2,560 feet (780 m) (see fig. 1.1). Still high up, the most impressive exposure of "rune stones" is encountered (fig. 1.5). Here, the mountain not only speaks but veritably sings its story to all who would pause to listen. From Bald Rock, the view back up the mountain is especially rewarding. Even at a leisurely pace, the descent returns the hiker to the paved highway by mid-afternoon. The day's hike has revealed much of what hides behind the mountain's three-part riddle, answering two questions of interest to geologists. Mount Monadnock stands tall because it is made of strong rock like quartzite. resistant to erosion by water and ice. The mountain's pedigree is further divulged by geological evidence showing how those metamorphic rocks were subject to deformities at great depths below the surface of the earth.

A UNIFYING CONCEPT IN PHYSICAL GEOGRAPHY

The last part of the riddle remains: What makes Mount Monadnock anything more than a single knob of real estate, the elevation of which is not all that impressive compared with other mountains? The significance is that the landscape surrounding Mount Monadnock was diminished by erosion to the extent that the resistant mountaintop is much higher by comparison. Although erosion by rivers considerably altered the New England landscape, those river

FIGURE 1.5 Lines scribbled like Norse runes in the mountain's rock announce the great depth of burial and pressure under which sillimanite minerals came to be altered as crystal morphs.

Photo by B. Gudveig Baarli.

valleys dissect the remnants of a flat surface called a peneplain that can easily be recognized from places like the Hoosac Range in western Massachusetts. William Morris Davis (1850–1934) was so struck by the binary relationship that the Harvard geographer wrote a paper published by the National Geographic Society in 1895 to define a monadnock as a universal concept. Thus, a monadnock is a conspicuous prominence that rises above a worn and equally conspicuous plain on account of differences in the susceptibility of rocks to erosion by natural agencies such as wind, rain, and ice.

Another prominent monadnock in North America is Black Elk Peak (formerly Harney Peak) in the Black Hills of South Dakota. Unlike Mount Monadnock, formed of quartzite, Black Elk Peak is formed of granite, and it resides at a higher elevation, 7,242 feet (2,207 m) but with a similar prominence of about 2,000 feet (610 m). Composed of resistant sandstone, Uluru (also known as Ayers Rock) is a world-famous monadnock located in central Australia. Its summit has an elevation of 2,831 feet (863 m) and a prominence of 1,142 feet (348 m) above the surrounding plains. Many present-day sky islands, such as Black Elk Peak and Uluru, have a prior history as genuine watery islands that reflect major changes in sea level through geologic time. A contemporary of William Morris Davis, Amadeus William Grabau (1870–1946) began his academic career at Columbia University but ended it in China, affiliated with Peking University. As a geologist, he thought deeply about the global record of sea-level change across time, and he coined his own name for a monadnock with a history as a functional island. In his book *The Rhythm of the Ages*, Grabau applied the Chinese place name for its northeastern province, Shantung, for just such a physical potentiality.[4]

STANDARD FOR DROWNED MONADNOCKS

Mount Monadnock is 65 miles (105 km) inland from the New England coast on the nearest approach from the North Atlantic. There is no evidence that seawater ever surrounded the mountain to make a genuine island. A sea-level rise of no less than 820 feet (250 m) above its present benchmark would be required to accomplish that end, something not about to happen anytime soon. In its stance as an iconic prominence, however, a flooded monadnock becomes a provisional island with outcrops that prefigure a rocky coastline. To visualize what Mount Monadnock might look like after long exposure to the work of stormy seas, it is instructive to find an example of a present-day drowned monadnock. It is a hunt that long consumed my thoughts. After rejecting several candidates, I found a superior example in Hongdo (Red Island) in the Yellow Sea, about 68 miles (110 km) off the southwest coast of South Korea.[5] Hongdo (fig. 1.6) is a Korean national monument formally incorporated within the Dadohae Marine National Park in 1981. It is a popular tourist attraction with Korean nationals, who throng to the island to enjoy summer boat tours esteemed for the red reflections of sunlight on coastal rock formations, especially during sunrise and sunset.

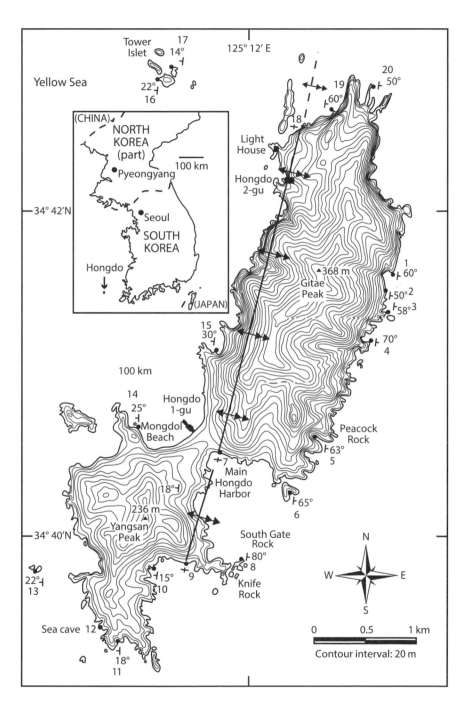

FIGURE 1.6 Elevations on Hongdo (Red Island) in the Yellow Sea off South Korea.

At nearly 1,700 acres (6.87 km²), Hongdo occupies one-third the map area of Mount Monadnock (compare fig. 1.1). The island is long and narrow in shape, but it fits perfectly within the same space as the Mount Monadnock precinct. Gitae Peak in the northern part of the island rises more than 1,200 feet (368 m) above sea level. Compared to Mount Monadnock, however, this altitude fails to consider the average water depth off the plunging rocky shores around the Korean island. The floor of the Yellow Sea around Hongdo sits at a water depth of about 260 feet (80 m). Hence, the island's elevation above the seabed amounts to 1,460 feet (448 m). Overall, the Korean island makes up in height above the seabed what it lacks in map area compared to Mount Monadnock. Even so, the botanical zonation so evident with altitude on Mount Monadnock is not matched by Hongdo, where the thin soil is covered by shrubs and small trees of the laurel (*Aucuba japonica*), silver magnolia (*Saussurea polylepis*), and Japanese camellia (*Camellia japonica*). Hongdo is justly renowned for its sea life, so evident in the diversity of marine invertebrates on sale at the wharf-side market. Sea cliffs exhibit a clear partition of biota through intertidal zones amounting to nearly 6.5 feet (2 m). The upper intertidal is dominated by the edible gooseneck barnacle (*Pollicipes mitella*), the middle part is crowded by the Far Eastern mussel (*Mytilus coruscus*), and the lower part shows a changeover to densely encrusting red algae.

Hongdo shares key similarities with Mount Monadnock regarding its rock composition and the structural deformation of bedded quartzite. The Korean island is a folded arch (an asymmetric anticline) shown by strike and dip values collected from a series of stations around the 13-mile (21-km) perimeter.[6] Quartzite layers on the east and west coasts dip away from each other in opposite directions, but those on the east are more than twice as steep as those on the west. In addition, the layers exposed on the east coast tend to be less thick than those on the opposite shore. As a result, the many islets scattered around the main island are smaller and closer to the shore on the east side and larger and farther offshore to the west because of controlled erosion. The named beauty spots and other coastal features around the island conform to a classic pattern of progressive erosion from sea caves to sea arches to isolated sea stacks. In the channel at Knife Rock off southern Hongdo, for example (fig. 1.7), it is possible to visit a slender sea stack represented by only a few layers of nearly vertical quartzite that stand as the surviving limb of an arch once connected to the mainland. No such sea arches or sea stacks are found around Mount Monadnock.

A continuous line drawn around the outer islets of Hongdo inscribes a former coastline with a larger perimeter, pointing to an earlier island of no less than 3,459 acres (14 km²). Based on the present core, the larger acreage suggests

FIGURE 1.7 Sea stack called the "Knife" formed by vertically tilted layers of quartzite on Hongdo, South Korea (author for scale).

Photo by B. Gudveig Baarli.

a loss of more than 50 percent in size over the last million years of Pleistocene time, when the Yellow Sea advanced and retreated multiple times. In effect, Red Island cycled back and forth between the statuses of real island and sky island. During intervals of high sea level, erosion was most effective on the west side of the island, where gently dipping layers of quartzite were vulnerable to the action of water forced under hydraulic pressure into partings and joints between beds of otherwise highly resistant quartzite. Today, Hongdo is subject to harsh winter squalls driven by cold fronts that arrive from the west on a biweekly basis. The island also sits in the path of subtropical typhoons that arrive from the southeast during the late summer and autumn. Marine erosion occurs on all sides of the island, although the dominant windward side is on the west coast. It is worth noting that Mongdol Beach is the only place anywhere around the island where eroded materials are accumulating as a cobble shingle (see fig. 1.6).

ISLAND DYNAMICS PRESENT AND PAST

Alone at sea, Hongdo is a speck of land well out of sight from the mainland Korean peninsula. Yet it embodies a compact island ecosystem in an isolated way station. A surprising amount of information about the island's physical construction and biological makeup may be collected from a small boat visiting every cove and corner of the shoreline over a few days under agreeable weather. For the geologist, it is possible to understand what kind of rocks the island is made from and how those rocks came to be exposed. For the student of geomorphology, it is possible to ascertain trends in coastal erosion related to windward and leeward settings on opposite shores affected by complex weather patterns. For the biologist, it is possible to tally the principal species that successfully colonized the island both above and below the mean tide line. For the ecologist, it is a manageable task to tease out the interactions of organisms and the physical world they occupy. It is an unfashionable notion nowadays, but the disciplinary roles of a geologist, geomorphologist, biologist, and ecologist may be combined and pursued under the guise of natural history. An untold number of small islands similar in size and complexity to the Korean Red Island are scattered throughout the world as sea-sculpted monadnocks, each in its own fashion capable of telling an intricate story about variations in physical form and biological accommodation.

Physical geographers and biogeographers divide the world's islands into three groups.[7] Continental islands, like Hongdo, are those surrounded by relatively shallow water on flooded continental shelves. The geology of such islands is as varied as the different rocks exposed on nearby continents. Oceanic islands include those isolated by deep water in midocean settings. Their geology is typically volcanic, often dominated by basalt. A third and more problematic category includes what has been called "ancient continental islands," a term often applied to larger islands like New Zealand and Cuba and smaller islands like Seychelles, formed from continental granite but adrift far out in the Indian Ocean. The global geology of plate tectonics, with its boundaries defined by rift and subduction zones, has sparked a reappraisal of how islands are categorized. For example, island arcs like the Aleutian Islands of Alaska, Japan, or western Indonesia (Sumatra and Java) owe their origin to convergent plate boundaries at subduction zones. Different grades of metamorphic rocks often are typical for these islands. Reclassification of midocean islands has resulted in subcategories that include intraplate hot spots (Hawaiian Islands), ocean-ridge hot spots (Iceland), and atoll hot spots (Diego Garcia), all associated with mantle-piercing plumes of magma mobilized from the mantle-core boundary located some 620 miles (1,000 km) deep within the earth. Such distinctions are nuanced, but the paleoislands described herein mostly denote simple continental-shelf islands comparable to present-day Hondo and simple midoceanic islands like those in the present-day Hawaiian chain of islands.

Paleoislands existed as active ecosystems during the deep past, millions, tens of millions, or even hundreds of millions of years ago. They are extinct islands, so to speak, with a history of burial or semiburial in the earth's crust, much the way envisioned by Grabau with his concept of shantungs.[8] Because such islands also are prone to be exhumed by erosion, the student of natural history can explore ancient shores and former ecosystems regardless of how old they might be. Using the Korean Red Island as a key to the past, various methods can be used to recognize a former rocky coastline based on evidence at geological junctions called unconformities.[9] Some proofs draw on physical geology, for example, the presence of conglomerate with pebbles or cobbles, much like those from Mongdol Beach on Hongdo. The conglomerate should be composed of the same parent rock found beneath. If multiple lithic clasts are present in the conglomerate, they should match the same kinds revealed in solid bedrock from place to place. The contact between underlying parent rocks and the overlying conglomerate may preserve a former wave-cut platform and even former tidal pools. An uneven contact with vertical relief may preserve abrupt sea cliffs and even former sea stacks. On a smaller scale, fissures that penetrate the parent rock from above may be filled with pebbles or sand contiguous with the overlying sedimentary deposit (neptunian dikes).

Other proofs draw on fossils potentially found at or near a geological unconformity.[10] Those like barnacles or oysters specialized for life in a setting with high to moderate energy from wave shock may be attached directly to the unconformity surface or to larger cobbles and boulders in the overlying deposit. Others might include clinging animals such as limpets, chitons, or other mollusks with a habit of wedging among boulders for stability in rough water. Coralline red algae also have the habit of forming thick crusts on rocks exposed to strong waves. Other fossils at or near the unconformity surface that reflect organisms normally associated with sheltered habitats in less agitated water, such as fragile Bryozoa, may be used to differentiate leeward from windward settings. Moreover, careful consideration of fossils from overlying sedimentary layers has the potential to show gradual changes in water depth as a monadnock is drowned by rising seas.

The idea that sea level fluctuates globally through geologic time is called eustasy.[11] It is a concept much debated as to its actual value. There is no dispute, for example, that sea level falls worldwide during episodes of major glaciation when huge volumes of water are put into cold storage on land in places like Greenland and Antarctica. Likewise, no one doubts that the sea level rises again with the melting of continental ice sheets. Because planet Earth is an imperfect spheroid, however, it is far from sure that sea level fluctuates to the same degree or even in perfect concert everywhere simultaneously. Geologists are more likely to speak of relative changes in sea level that take into account factors including local and regional tectonics.

Figure 1.8 covers possible outcomes for island monadnocks in response to changes in sea level. A major eustatic rise or fall at the same time that an island is statically fixed in place (fig. 1.8a) will result in significant exposure or flooding in response to events like the passage of ice ages. However, local uplift during the same time when the eustatic sea remains neutral (fig. 1.8b) yields a substantial increase in island exposure. Local tectonic uplift of an island during a major eustatic rise or fall (fig. 1.8c) leads to either a large change in relative position or no change at all when contrary actions cancel out each other. Minor changes in sea level at the same time as large-scale tectonic subsidence (fig. 1.8d) always result in the complete foundering of the island. The other

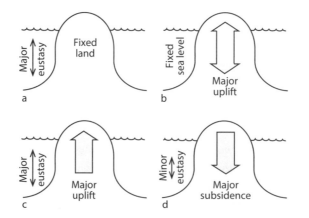

FIGURE 1.8 Interpretive models for changes in relative sea level: a) eustatic rise or fall of sea level while island is fixed statically in place; b) fixed sea level while island is subject to major tectonic uplift; c) island is subject to major local uplift while sea level undergoes a major eustatic rise or fall; d) island is subject of major tectonic subsidence during minor eustatic changes in sea level.

three scenarios can potentially leave an island "dry-docked" on land so that its structure and biological artifacts may be studied both above and below the former water line.

ITINERARY FOR AN IMMENSE JOURNEY

On climbing Mount Monadnock, the view from the summit reaches far beyond the surrounding New Hampshire countryside to encompass seascapes populated by ancient islands scattered widely both in geographic space and geologic time. The iconic status of a monadnock as a geographic prominence resistant to change by powerful agents of erosion over deep time derives from this very spot as bestowed by William Morris Davis in 1895. The Mount Monadnock label is scientific in context and carries a sense of reverence that descends from the transcendentalist tradition invoked by Ralph Waldo Emerson and his followers in the mid-1800s. Our blue planet has a vast history, and the shores of stony islands retain a record of biological change in compact ecosystems that survived persistent sea swells driven by prevailing winds, recurring sea storms, changing sea levels, and local tectonics. Such a history is worthy of reverence by those who have inherited the planet and have the intellectual capabilities to act responsibly as its stewards.

To trace such a long and eventful history and examine fossil evidence from jellyfish to coral reefs to swimming dinosaurs is to undertake an immense journey. Our itinerary involves time travel to visit not only particular paleoislands but also whole archipelagos as laid out in figure 1.9, a time chart for the last half billion years. The multitude of names for intervals of time long past is unwieldy, but the burden can be shouldered as a necessary chart that guides our way to a succession of planets very different from the one we now occupy. In the ensuing chapters, Paleozoic pathways from about 500 to 250 million years ago are traced in and around the Cambrian Baraboo Archipelago of Wisconsin (United States), the Ordovician Jens Munk Archipelago of Manitoba (Canada), Silurian Bater Island of Inner Mongolia (China), as well as the Devonian Mowanbini Archipelago on the Oscar Range and the Permian karst-island labyrinth near Gascoyne Junction, both in Western Australia. Mesozoic pathways from about 199 to 65 million years are followed among islands of the Saint David Archipelago (South Wales) and Cretaceous islands of the Pacific-coast Erendira Archipelago (Mexico). Cenozoic pathways that bridge the last fifteen million years attend to oceanic islands from the Miocene

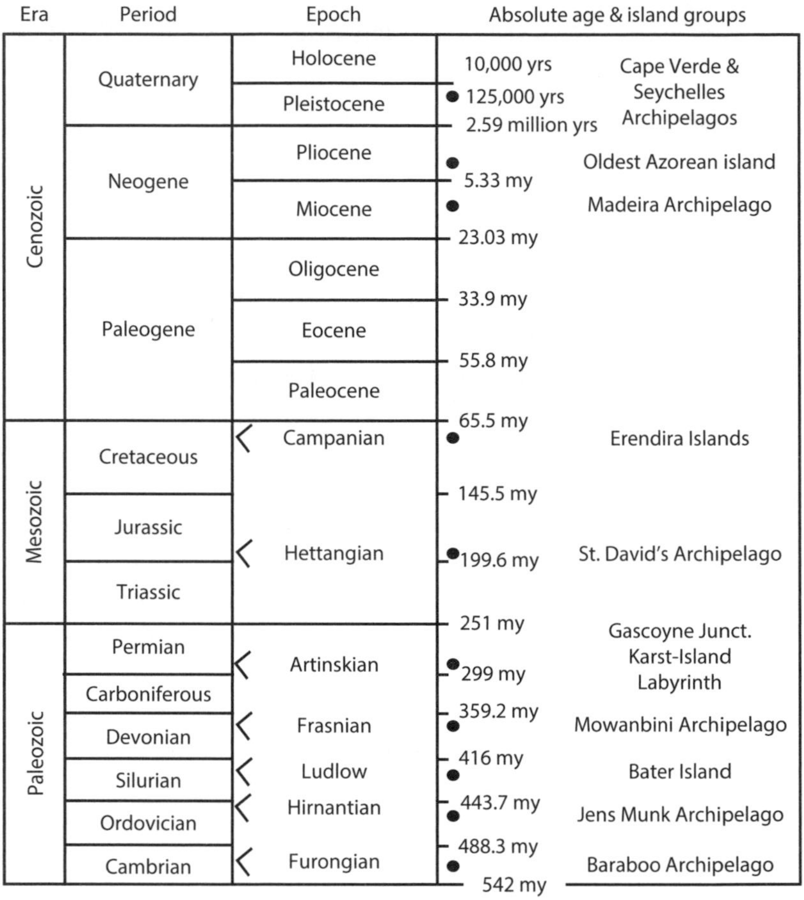

Era	Period	Epoch	Absolute age & island groups	
Cenozoic	Quaternary	Holocene	10,000 yrs	Cape Verde & Seychelles Archipelagos
Cenozoic	Quaternary	Pleistocene	● 125,000 yrs	Cape Verde & Seychelles Archipelagos
Cenozoic	Quaternary	Pleistocene	2.59 million yrs	Cape Verde & Seychelles Archipelagos
Cenozoic	Neogene	Pliocene	●	Oldest Azorean island
Cenozoic	Neogene	Pliocene	5.33 my	Oldest Azorean island
Cenozoic	Neogene	Miocene	●	Madeira Archipelago
Cenozoic	Neogene	Miocene	23.03 my	Madeira Archipelago
Cenozoic	Paleogene	Oligocene	33.9 my	
Cenozoic	Paleogene	Eocene	55.8 my	
Cenozoic	Paleogene	Paleocene	65.5 my	
Mesozoic	Cretaceous	Campanian	●	Erendira Islands
Mesozoic	Cretaceous	Campanian	145.5 my	Erendira Islands
Mesozoic	Jurassic	Hettangian	●199.6 my	St. David's Archipelago
Mesozoic	Triassic	Hettangian	251 my	
Paleozoic	Permian	Artinskian	●299 my	Gascoyne Junct. Karst-Island Labyrinth
Paleozoic	Carboniferous	Artinskian	359.2 my	Gascoyne Junct. Karst-Island Labyrinth
Paleozoic	Devonian	Frasnian	●	Mowanbini Archipelago
Paleozoic	Silurian	Ludlow	416 my ●	Bater Island
Paleozoic	Ordovician	Hirnantian	443.7 my ●	Jens Munk Archipelago
Paleozoic	Cambrian	Furongian	488.3 my ●	Baraboo Archipelago
			542 my	

FIGURE 1.9 Geologic time scale denoting the age of different paleoislands and former archipelagos covered in this volume.

to Pleistocene Madeira and Azorean Archipelagos; Cape Verde Archipelago (all in the North Atlantic Ocean); Seychelles (Indian Ocean); and Hawaiian (Pacific Ocean) Archipelagos. Our immense journey concludes with thoughts on islands as we find them today and hope to see them conserved under human stewardship in the future.

CHAPTER 2

HOW AN ISLAND CLUSTER ACQUIRES ITS SHAPE

A Journey in Late Cambrian Time to Wisconsin's Baraboo Archipelago

When we go down to the low-tide line, we enter a world that is as old as the Earth itself—the primeval meeting place of the elements of earth and water.

—Rachel Carson, *The Edge of the Sea* (1955)

Natural Bridge State Park in North Adams, Massachusetts, celebrates the preservation of a white marble archway spanning a gorge cut through bedrock 60 feet (18 m) deep on Hudson Brook. The marble is a metamorphic rock that resulted from the deformation of Cambrian limestone altered by heat and pressure after deep burial during the Taconic orogeny in later Ordovician time. Close to my home in the Berkshires of western Massachusetts, the park's orderly grounds offer a respite from humid summer days with cool air rising from the tumult of waters that splash through the twisted, narrow chasm below the arch. A nearby bench offers a bird's-eye view of an abandoned marble quarry that began operations during the early 1800s. The quarry was the principal source of tombstones, many of which stand today in New England burial grounds where some who witnessed the American Revolution more than two centuries ago now rest in peace. The marble's original lime content was deposited on the floor of a shallow marine shelf more than five hundred million years ago during the Cambrian period (see time chart, fig. 1.9). Five hundred million years is a span roughly 2.5 million times longer than the length of time the oldest grave markers have stood undisturbed as simple but elegant memorials hewn from the big quarry.

During the latter part of the Cambrian the level of the sea stood much higher than today, and much of the ancestral North American continent was awash in salt water. Named by geologists as Laurentia, after the Laurentides of southern Quebec at the edge of the Canadian shield, the ancestral continent was smaller than the North America we know today. From the park bench on the outskirts of North Adams, it takes a special kind of geological insight to appreciate that the same spot five hundred million years ago was below sea level, perched near the edge of a continental shelf. Regarding today's geography, the nearest landfall was to the northwest on the opposite side of the Saint Laurence River in Quebec. Westward from Massachusetts following a present-day latitude around 43° north of the equator, the distance to the nearest islands on the Cambrian continent is 835 miles (1,345 km) in Baraboo, Wisconsin (fig. 2.1a). After Baraboo, the next recognizable Cambrian islands are 670 miles (1,080 km) away near Rapid City in the Black Hills of South Dakota. A string of larger Cambrian islands across today's Dakotas, Nebraska, Kansas, Oklahoma, and Texas are linked together on the Cambrian Transcontinental Arch through the center of the ancient continent.

This chapter invites a journey to the Cambrian Baraboo Archipelago of Wisconsin (fig. 2.1b) with its cluster of about thirty individual monadnocks akin to the quartzite Mount Monadnock of New Hampshire. Extra coverage includes similar islands of coeval Cambrian age in the Black Hills of South Dakota and the Slick Hills of southwestern Oklahoma. Centered around Devil's Lake, few of the monadnocks in the Baraboo district are larger than the area occupied by Mount Monadnock itself. Half the Baraboo monadnocks are smaller, on average less than 1,000 acres (4 km²) in size, or comparable to only 20 percent of the uplands around Mount Monadnock. A combination geologic and topographic map (fig. 2.2) covers the district west of U.S. Highway 12 and features the largest Baraboo monadnock with an area of 20 square miles (~ 50 km²), two and a half times the area of Mount Monadnock. Whereas the prominence of Mount Monadnock amounts to 2,000 feet (610 m), the largest Baraboo monadnock has a maximum prominence of scarcely 570 feet (174 m) above the surrounding Wisconsin farmland. The halo of eroded Cambrian conglomerate around the perimeter of the big monadnock speaks to the fact that the Cambrian sea level left a distinct mark equal, on average, to an elevation of 1,148 feet (350 m) above present-day sea level. With sea level as a benchmark for comparison, it means that the big monadnock expresses only half the prominence of today's Hongdo (Red Island) in the Yellow Sea off South Korea (see fig. 1.6). This assessment,

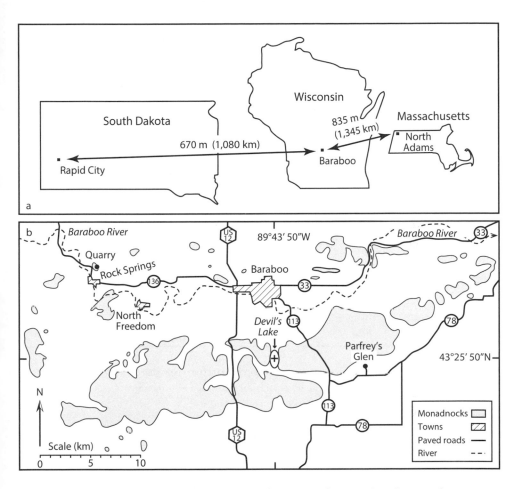

FIGURE 2.1 Introduction to the Baraboo area of Wisconsin: a) State outlines for Massachusetts, Wisconsin, and South Dakota showing relative distances between points of comparison; b) Pattern of quartzite monadnocks clustered around the town of Baraboo in southern Wisconsin.

however, fails to acknowledge the long history of potential erosion off the top of the big Baraboo monadnock over the last five hundred million years. Even so, the distinctive halo of surviving Cambrian conglomerate defines the lasting shape of the individual Baraboo islands, both large and small. The aim of this journey is to understand how such a cluster of ancient islands acquired its shape and distribution, surviving to give us so clear a portraiture of that most primeval meeting place between the elements of earth and water to which Rachel Carson so eloquently alludes.

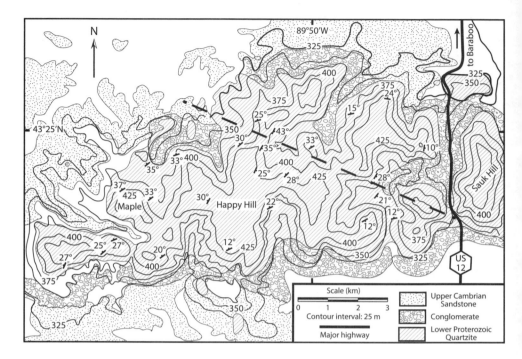

FIGURE 2.2 Coverage representing the largest quartzite monadnock and surrounding Cambrian rock formations with conglomerate and sandstone near Baraboo, Wisconsin (note position of transect line A–B). Modified from Dalziel and Dott, 1970.

FEET ON THE GROUND

An accurate topographic map paired with a geologic map offers a powerful introduction to any given place, no less so than for the largest monadnock in the Baraboo region (fig. 2.2). Stepping onto the shore of this paleoisland, we must take into account that the Cambrian period had nearly come to a close after a run of fifty-four million years. During that time, the diversity of marine invertebrates proliferated to include a wide range of organisms represented by nearly all the major marine invertebrate groups today. Land plants did not yet exist in this world. Our geologic clock is now set five hundred million years ago. We are confronted by a paleoisland that is utterly devoid of vegetation. Map in hand, it is apparent that concentric lines of equal elevation encircle Happy Hill near the center of the complex, not unlike the pattern of Mount Monadnock (see fig. 1.1). Contrary to Mount Monadnock, however, the big Baraboo monadnock exhibits a crenulated topography with radial valleys filled by Cambrian conglomerate that penetrate as much as a mile (3 km) toward the center from the periphery. The modest V-shaped indentations on topographic lines crossed by rivulets on the flanks of Mount Monadnock are nothing like the deeply entrenched valleys eroded around the margin of the monadnock during late Cambrian time. Comparison with the Korean Red Island is instructive on account of that island's opposing bays, one of which is filled with eroded quartzite cobbles at Mongdol Beach (see fig. 1.6).

Additional insights are hard won only through direct exploration of the monadnock on foot, following a course for some 6 miles (10 km) on a compass bearing N 67° W (line A–B on fig. 2.2). Few secondary roads cross the area. Today, the monadnock is thickly forested, and a compass is necessary to stay true to the bearing. The starting point is from an elevation 1,066 feet (325 m) above sea level at the bend in U.S. 12 near the southeast corner of the monadnock. It takes an initial climb of 330 feet (100 m) to reach the summit of the first hillock three-quarters of a mile (1.2 km) inland. The slope rises on a 10 percent grade, and it affords places to stop and examine monoliths of quartzite that stand erect in the forest, not unlike ghostly sea stacks on a wave-swept coast (fig. 2.3). Unsurprisingly, the Baraboo quartzite is the name for the monadnock's core rock. Unlike the younger quartzite in the Littleton Formation of Mount Monadnock, which is gray in tone and indicative of clean silica sand, the Baraboo quartzite is purple to maroon in color due to iron impurities that stained the original quartz grains and colored the cement that binds those grains together.

The crest of the first hillock is oval-shaped, with an orientation parallel to the transect we follow. Crossing level terrain for the next half mile (800 m), there are no useful exposures of quartzite. Besides stone quarries and road cuttings, few places in the district offer natural outcrops of any considerable thickness. Even so, what geologists have learned about the Baraboo quartzite offers much to consider. The maximum thickness of layered quartzite amounts to something close to 3,600 feet (as much as 1,100 m), and the age of the original sandstone subsequently deformed by metamorphism dates to 1.7 billion years at the close of the Paleoproterozoic.[1] Generally, the lower part of the package consists of river sand that once accumulated in a braided stream. Thin lenses of pebble conglomerate are interspersed among sandy layers. The upper part of the formation was influenced by tides in a shallow marine setting distinguished by sedimentary structures that include tabular cross-stratification. Iron staining by the mineral hematite helps to make the layering more conspicuous. Ripple pavements are well developed at horizons in the upper part of the succession. The preserved asymmetry of ripples and dip orientations from cross-bedding suggest that tidal

FIGURE 2.3 Quartzite pinnacle near US Highway 12 on the slope of the region's largest monadnock.

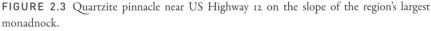

Photo by author.

energy shifted sediments from the present northwest to the southeast during deposition in the distant Paleoproterozoic long before the first Cambrian sediments were eroded from the hardened monadnocks transformed by metamorphism. In effect, the present-day landscape of the Baraboo district is an enigma that presents itself as a sequence of nested boxes that must be unpacked in the right order to reveal the area's complete natural history. The original sediments from the Baraboo quartzite constitute that most inner, hidden box that gives us the starting story.

The next mile (1.6 km) along the transect ends at a ravine slightly below the same line of elevation where we began but on the far side of the hillock. Here, the streambed cuts through conglomerate formed by quartzite cobbles eroded from the monadnock during the late Cambrian. We have crossed a boundary (unconformity) between Cambrian strata close to five hundred million years old and the underlying 1.7 billion-year-old parent quartzite from which the conglomerate was derived. A former Cambrian valley descends eastward to intersect the main artery through which the north–south highway runs (fig. 2.2). Geologists from the University of Wisconsin in Madison devoted much effort to measuring the size and relative smoothness of cobbles and small boulders within the Cambrian conglomerate that rings the many monadnocks in the Baraboo district.[2] Those Cambrian clasts are mostly smooth and well-rounded, but they range across a full array of sizes from pebbles to boulders as much as 6.5 feet (2 m) in diameter. Here, the cobble-size quartzite clasts are angular in shape. The implication is that wave action was sufficient to loosen material from the parent rock at the head of the valley but insufficient to smooth the rough edges over time through the jostle of clasts against one another.

Climbing out of the ravine, we reach a quartzite outcrop where it is possible to measure the inclination of bedding planes that dip 28° to the southeast. Our line of march continues for more than a half mile (1 km) to reach a saddle at an elevation of about 1,345 feet (410 m) between two hillocks rising on either side of the transect. The topographic saddle denotes a divide on the monadnock from which today's drainage flows in opposite directions. Farther along, another quartzite outcrop exposes layers that dip 33° to the northwest (fig. 2.2). In contrast to the earlier measurement, it means that the topographic saddle and the hillocks it connects represent the crest of a small, folded arch (anticline) deformed in the original Baraboo quartzite. Bedding from divergent limbs of the arch slope away from each other in opposite directions.

Outcrops farther along the transect over the next mile and a half (2 km) reveal yet another fold, much like the previous example, also related to a topographic saddle. In this case, beds within the quartzite dip more steeply 43° on one side to the southeast and less steeply 25° to the northwest on the other. The ground between these two folds where the dipping limbs converge denotes a structural trough (syncline). In scale, it helps to think about a stack of paper piled with fifty sheets, one atop another, each sheet representing a rock layer. If two opposite ends of the pile are pushed toward each other, the wad of paper buckles into a series of parallel folds separated by crests and valleys. Like the bended knee of Mount Monadnock (see chapter 1), such compression occurs at a great depth below the surface where the hardest rocks behave like toffee candy when hot and malleable.

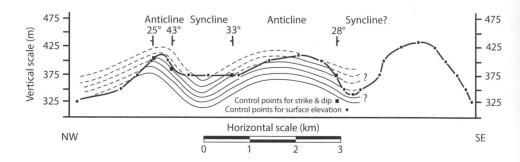

FIGURE 2.4 Cross-section panel based on interpretation of strike and dip values encountered on the transect A–B (from topo map 2.2). Dashed lines project the extension of former quartzite layers above the present surface that were removed by erosion.

Using a piece of graph paper, the total of the topographic information based on the measured dips may be combined to plot a cross-section cut by the transect across Happy Hill (fig. 2.4). The cross-section tells its own story, like that revealed by opening a hidden box. It is a narrative about what happened to the original sandy sediments as they were transformed into quartzite and subjected to the compressive forces that led to folding. Any number of transects across other parts of the same monadnock or other smaller monadnocks in the Baraboo area can be expected to yield a more refined picture of the physical forces that shaped the islands and caused them to reach above the Cambrian seascape. The end of the transect (fig. 2.2) leads downslope across the boundary between the Baraboo quartzite and adjoining Cambrian sandstone. Sandstone is exposed in ravines where pebbles and cobbles of eroded quartzite are embedded, but more angular boulders are entrained nearby.

OTHER PARTS OF THE ARCHIPELAGO

At least two additional stops yield a critical overview of sites in the Cambrian Archipelago that are especially informative. The first is at Parfrey's Glen, located 10 miles (16 km) east of the center of the big monadnock. It is reached by a side road off Wisconsin Highway 113 from Baraboo to the southeast (fig 2.1b). The glen is a deep ravine within Devil's Lake State Park that was awarded protection as a State Natural Area in 1952. The place holds special meaning to me because I visited as a teenager and experienced early stirrings of interest in what would become a life-long career in geology.[3]

From the parking area at the end of an access road, a half-mile (0.75 km) trail leads northward to follow an entrenched steam bed. In various states of repair during its history as a protected site, the trail featured boardwalks and stairs to a viewing platform that aided the visitor in fording the stream and getting a better look at canyon walls. Recently, powerful floods degraded the infrastructure to the point where repairs were no longer deemed practical. On a clear summer's day, however, hiking the streambed in rubber boots makes a pleasant excursion through the cool glen shaded by yellow birch (*Betula alleghaniensis*) and mountain maple (*Acer spicatum*). Dense growth by moss and ferns on the lower walls of the narrowing ravine was cleared away during an especially violent flood in 2010. The violence of that particular storm cleaned rock walls that feature thick beds of conglomerate and interbedded sandstone. Now, as millions of years ago, powerful storms represent a veritable force of nature. Quartzite clasts packed in the conglomerate and scattered through the interbedded sandstone layers, all derived from the wave-swept rocky shores on a nearby monadnock.

Strata within the 50-foot (15 m) cliff face near the trail's end are conveyed by a sketch (fig. 2.5) showing the repetition of multiple storm events through a few million years. The succession of individual layers, some with more or less conglomerate, is formally attributed to the upper part of a geological unit called the Tunnel City Group. Each layer denotes a particular event in ecologic time. Some sand layers are distinguished by cross-bedding. High above, a thick bed

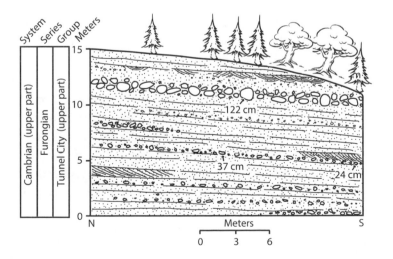

FIGURE 2.5 East face of Parfreys Glen (see fig. 2.1b for location), showing thick succession of sandstone and conglomerate layers in the Cambrian Tunnel City Group eroded from the adjacent quartzite monadnock to the north. Modified from fig. 40 in Dalziel and Dott, 1970.

of conglomerate contains quartzite boulders up to 4 feet (122 cm) in diameter, held in place by a sandy matrix (fig. 2.5). The storm surge during that event must have been extraordinary to carry off such large and heavy boulders from nearby rocky shores. Sand layers without cross-bedding denote intervals of time under weather conditions when normal water circulation was capable of depositing finer-grained sediment with ease. Cross-bedding in other sandstone beds denotes a higher regimen of water energy under which the transport of materials across the seabed clearly was directional.

The Wisconsin Tunnel City Group is correlated with the Furongian series, which occupies the uppermost part of the global Cambrian system (figs. 1.9 and 2.5). The corresponding Furongian epoch is a well-defined interval of geological time that lasted more than ten million years until the close of the Cambrian period, about 488 million years ago. Under the authority of the International Union of Geological Sciences, all igneous rocks and sedimentary strata formed during that interval of geologic time anywhere in the world are regarded as belonging to the Furongian series.[4] Although Cambrian trilobite fossils from Wisconsin lack a global distribution, the onset of a positive shift in carbon-13 isotopes recognized worldwide as the Steptoean Positive Carbonate Isotope Excursion (SPICE) is well represented in North America. Technically, this is significant because the substantial thickness of conglomerate and sandstone deposited as halos around the margins of the many Baraboo quartzite monadnocks can be correlated with geological formations with the same trace isotopes all around the world. Even so, few other regions with rocks of this age present such a refined picture of island development.

An abandoned quarry near Rock Springs is the second site worthy of an extended visit in the Baraboo district. It is accessible on Wisconsin State Highway 136, 9 miles (15 km) west of Baraboo (fig. 2.1b). A corner of the quarry (fig. 2.6) is excavated to a depth of about 40 feet (12 m) below the surface of the surrounding Wisconsin countryside. Most of the exposed rock within the quarry belongs to the Baraboo quartzite, which reveals thick layers standing vertically upright on end. The quarry is quite large, with a north–south extent of nearly 1,000 feet (~300 m), which makes it easy to directly measure thickness along the quarry floor perpendicular to the strike of the quartzite beds. A deep trough (syncline) is interpreted on the assumption that vertically oriented beds dive deep into the ground but bend back toward the surface at considerable depth to join with layers from the southerly monadnocks like Happy Hill that dip more gently northward.[5]

Effectively, the region's overriding geological structure is shaped like a huge trough extending more or less east to west in orientation but with sides that are

FIGURE 2.6 Abandoned quarry north of Rock Springs (see fig. 2.1b for location), showing burial of a quartzite monadnock by a thin cover of Cambrian sandstone.

Photo by author.

asymmetrical in cross-section. One side is vertical, but the other is more like a gentle ramp descending to intersect the steeper limb at great depth. Smaller monadnock islands distributed across the northern part of the Baraboo district are those formed by outcrops with vertical or steeply dipping quartzite layers. In contrast, the larger monadnock islands to the south are derived from the ramp-like part of the trough. Such a regional analysis does nothing to invalidate the cross-sectional view of smaller anticlines and synclines detected on the transect across the largest of the southern monadnocks (fig. 2.4). The trend of those smaller structures associated with the topography of the big monadnock merely denotes wrinkles spread crosswise on the larger ramp. In total, these variations in the dip and orientation of quartzite layers fundamentally control the size and shape of each island in the Baraboo Archipelago and its ring-like continence.

The Rock Springs quarry also shows the gradual burial of a small Cambrian island during Furongian time. On the right side of the quarry photo (fig. 2.6), it is observed that strata in the Tunnel City Group rise against the side of the Baraboo quartzite, intersecting its layers across a slope. Those sedimentary strata thin out at the top of the quarry (middle to left side of the photo), formerly the top of an islet. The relationship is familiar to geologists as an angular unconformity. That is to say, younger strata sit atop older strata that are no longer conformable to one another and reflect different orientations in their bedding. The physical unconformity posits a time gap in the rock record between the stage when the older strata were structurally deformed by uplift and when the younger strata were deposited above. As we already know, the gap in time was exceedingly great, amounting to more than a billion years between the Late Paleoproterozoic and Late Cambrian.

The quarry locality provides additional insight beyond what is observed at Parfrey's Glen. At the glen (fig. 2.5), we cannot see the contact between conglomeratic beds from the upper part of the Tunnel City Group and a nearby rocky shore. However, the contents of those Furongian strata must have derived from coastal erosion. Here at the quarry, the relationship is definitive. Once the smaller monadnock island was fully buried, it was no longer available as a source to feed eroded materials into the shallow Cambrian sea. Likewise, the immediate juxtaposition of Tunnel City strata against the Baraboo quartzite at the quarry site tells us that a relative rise in the Cambrian sea level occurred. The present-day elevation at the top of the Rock Springs quarry is roughly 360 feet (~110 m) lower than the summit of the island monadnock nearest Parfrey's Glen, which means that some island monadnocks in the Baraboo area were susceptible to burial well before other larger islands in the same archipelago.

Moreover, it can be argued that the vertical arrangement of quartzite layers typical for the smaller northern islands made them more vulnerable to coastal attack by wave surge than the larger southern islands with the same kind of quartzite but with ramped layers in a configuration that repelled the force of waves more effectively. Based on surviving Ordovician strata also in contact with Baraboo quartzite on the western flank of the big monadnock (blank spaces in fig. 2.2), it is certain that the larger islands persisted as geographic features into the early Ordovician period. A relative rise in sea level no less than 50 feet (15 m) is confirmed by relationships found at Parfrey's Glen (fig. 2.5) and the Rock Springs quarry (fig. 2.6), but whether or not the sea level rose or fell in unison elsewhere around the Cambrian world is less certain.

STORM SIGNALS

Investigations on the sedimentology of rock formations in the Furongian series of south-central and eastern Wisconsin place that region in the context of a Cambrian continent with a much different orientation than today's North America. The geologist Robert Dott Jr. put the Wisconsin Baraboo district at a latitude south of the Cambrian equator and championed the idea that the accumulation of thick conglomerates in the Tunnel City Group and succeeding rock units was the product of tropical storms lashing against increasingly drowned monadnock islands.[6] Subsequent studies by Jennifer Eoff updated correlations of coeval formations throughout the Upper Mississippi River Valley, paying attention to variations in cross-bedded sandstone, including sedimentary structures often related to storm activity in modern settings.[7] The ancestral location of southern Wisconsin puts the region at a latitude about 20° south of the Cambrian equator. This placement is in agreement with the argument by Dott that trade winds had a direct impact on what is today the northern fringe of the Baraboo Archipelago.

Given sufficient fetch over open water, persistent trade winds can set up swells with crests far apart but commonly on the order of 8 feet (2.5 m) from crest to crest in deep water. Due to frictional drag with the seabed, this energy tends to be rapidly dissipated as ocean swells cross shallow water onto continental shelves or the shelves around oceanic islands. Theoretically, because the Laurentian continent during Cambrian time was flooded by shallow seawater and the Baraboo Archipelago was so far from the continental shelf, little wave energy would be left to expend on the rocky shores of the Baraboo monadnocks by the time wind-driven waves arrived. Such reasoning discounts evidence for the extensive Tunnel City conglomerate found around most of the island monadnocks in the Baraboo district. Given the upper size range of quartzite boulders deposited in conglomerate beds or dispersed within sandy layers, the physical reality of big storm waves cannot be ignored. Recent analysis argues that producing subspherical quartzite boulders with a diameter of 5 feet (1.5 m) would require extensive tumbling by waves with breaker heights of 23 to 26 feet (7–8 m).[8] Moreover, it is noted that quartzite boulders typically exhibit percussion marks, or small crescent-shaped fractures that result from impact with other boulders.

Wave dynamics related to today's trade winds account for extensive coastal erosion on islands like South Korea's Hongdo in relatively shallow-water settings like the Yellow Sea, where winter storms are a factor associated with the strong westerly flow of winds. Other forces alluded to as tropical storms must be invoked to account for the overall pattern of erosion in the Cambrian Baraboo Archipelago.[9] Data on the distribution and size range of eroded quartzite clasts in strata of the Tunnel City Group show the futility of detecting any pattern of windward and leeward subenvironments within the archipelago. But the distribution of boulders with more angular shapes, as opposed to those with subrounded shapes, might suggest such a dichotomy on the assumption that irregular boulders suffered less agitation with wave energy in more sheltered, leeward settings.

As found on the transect across the big monadnock (fig. 2.2), the occurrence of irregular boulders can be explained by their location within flooded valleys that led well inland where wave energy should have been dissipated. However, this reasoning cannot explain the wider distribution of irregular boulders. An alternative explanation for the largest irregular boulders is simply that they are too big to be regularly overturned by even the largest waves. At the upper end of an energy continuum, tropical and subtropical storms of hurricane intensity are the most logical explanation for the general pattern of quartzite clast distribution in the Upper Cambrian strata of the Baraboo district. Defined by threshold wind speeds on the Saffir-Simpson scale ranging from 74 to 156 miles per hour (119 to 252 km/h), there is no reason to doubt that storms equal to between category 1 and 5 hurricanes visited the Baraboo Archipelago.

HISTORICAL VIGNETTE: CHRISTINA LOCHMAN-BALK ON THE CAMBRIAN CRATON

Christina Lochman-Balk (1907–2006) led a distinguished career as a paleontologist, biostratigrapher, and paleogeographer specializing in the Upper Cambrian system in North America.[10] Foremost among her achievements was the publication of a detailed set of seven paleogeographic maps covering much of the lower forty-eight states in the United States and parts of adjacent Canada and Mexico.[11] The biostratigraphy that defines shifting timelines behind each of the maps was based on her extensive knowledge of Cambrian trilobites, which she organized into discrete zones tracking the growth and demise of related faunas through their initial and final appearances as fossils in Cambrian strata at more than 250 outcrops and subsurface cuttings. The resulting maps show cyclic changes in the expansion and reduction of lands on the Cambrian continent over what has been called the Great Unconformity, stretching between present-day Wisconsin and the Grand Canyon of Arizona. During sea-level rises, the region went from broad swaths of land covering a great expanse of today's midwestern and some western states to a line of islands along the continent's spine. The reverse took place during the retreat of seas toward the continental shelves. Lochman-Balk laid out convincing evidence for three episodes of rising sea levels among four episodes of falling sea levels depicted by the sequence in her seven paleogeographic maps.

The penultimate reconstruction in the map series represents a record-high stand in sea level commensurate with a unique biostratigraphic zone defined by more than forty trilobite genera. This particular map from Lochman-Balk

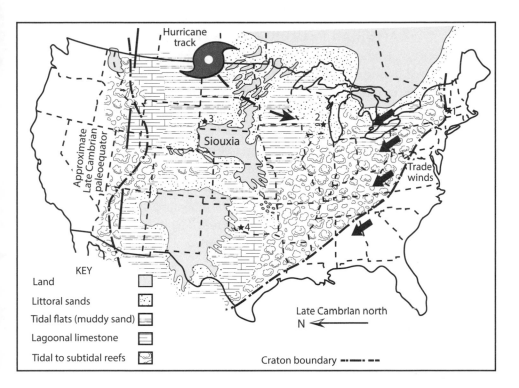

FIGURE 2.7 Paleogeographic interpretation of the ancestral North American continent at a record high stand in sea level during the latest Furongian epoch of the Cambrian period. Modified from fig. 6 in Lochman-Balk (1970) adding Cambrian equator, trade winds, and hurricane track. Localities denoted on the map include present-day western Massachusetts (1), the Baraboo district of Wisconsin (2), Black Hills of South Dakota (3), and Slick Hills of southwestern Oklahoma (4).

corresponds to a timeline in the Furongian epoch as subsequently designated in 2003 and reproduced here with modifications (fig. 2.7). As late as 1970, historical geology had yet to fully embrace the revolution in plate tectonics. Based on geomagnetic studies unfamiliar to Lockman-Balk and better refined later on, the Cambrian continent is interpreted to have occupied a position 90° clockwise from the present orientation of North America. Hence, the updated reconstruction in figure 2.7 shows the approximate paleoequator during the Late Cambrian. It also invokes the flow of persistent trade winds toward the equator and a deviation away from the equator on a hurricane track typical for latitudes between 10° and 20° of latitude south of the equator.

CAMBRIAN PELAGIC AND INTERTIDAL LIFE

Life thrived during the Cambrian in what is now Wisconsin, but it was dominated mostly by trilobites that proliferated in shallow, subtidal settings. Jellyfish or scyphozoan medusa represent the most unusual and unexpected discovery among other Cambrian marine invertebrates.[12] Scores of individuals, some as large as 20 inches (50 cm) in diameter (fig. 2.8), occur as impressions on bedding surfaces of rippled sandstone in formations directly preceding the Tunnel City Group but farther north from Baraboo in central Wisconsin. In many cases, preservation is sufficient to show the details of subumbrellar margins and gastrovascular cavities comparable to today's jellyfish. The fossil jellyfish are interpreted as mass stranding events that brought swarms onto tidal flats due to winds and currents.

However, close attention to quartzite surfaces on former rocky shorelines in the Baraboo district has failed to reveal evidence for incrusting or clinging

FIGURE 2.8 Bedding surface with impressions of Cambrian medusa (jellyfish) exposed in the Krukowski quarry near Mosinee in central Wisconsin (broom for scale).

Photo: James W. Hagadorn.

organisms of the kind typically adapted to rough-water settings today. The same applies to the abundant cobbles and boulders in the conglomerate of the Tunnel City Group plastered against the Cambrian monadnocks. Little evidence of fossil material has been detected, even within the sandy matrix among the conglomeratic clasts. We know that a diverse intertidal biota is accommodated by quartzite surfaces because today, such life is plentiful in places like South Korea's Red Island. But intertidal life was not as fully evolved during the Cambrian as now.[13] The intensity of wave energy around the Baraboo monadnocks may have been too violent to harbor life in such a niche. However, a more careful search is necessary to prove otherwise.

OTHER CAMBRIAN ISLANDS

Leaving behind the Baraboo Hills and advancing on the same line of exploration from the Cambrian shelf margin in Western Massachusetts (fig. 2.1a), the distance is 670 miles (1,080 km) to Rapid City, South Dakota. The city is the gateway to the Black Hills, the site of another sky island at Black Elk Peak (see chapter 1). Between Baraboo and Rapid City, the flat farmlands of Iowa and adjacent South Dakota conceal an expanse of Paleoproterozoic rock called the Sioux quartzite. Outcrops poke above the farmlands in northwestern Iowa and especially neighboring South Dakota along the Big Sioux River, where the strikingly pink rock is well exposed. Rebuilt after a devastating fire in 1888, the main street of the small South Dakota town of Dell Rapids is lined with handsome buildings quarried from the local quartzite, including a hotel, bank, and opera house. Mostly out of sight below the surface, the Sioux quartzite may have contributed to islands of its own making.

Lochman-Balk conducted fieldwork in South Dakota focusing on the Upper Cambrian Deadwood formation, which she placed at the edge of a geographic high called Siouxia (fig. 2.7).[14] She described local erosion on the northeast rim of the Black Hills during the Late Cambrian as shedding quartzite conglomerate to a maximum thickness of 50 feet (~15 m) against the flank of a rocky quartzite shore. Like the conglomerates of the Tunnel City Group in Wisconsin, those in the Black Hills are related to a rise in sea level during the Furongian epoch. Near Norris Peak on the east side of the Black Hills, quartzite boulders from the base of the Deadwood Formation rest on near-vertical beds of Paleoproterozoic quartzite that constitute the direct source of surf-eroded boulders. Among these, a boulder more than 2 feet (60 cm) in diameter reveals evidence of borings in quartzite made by an unknown Cambrian organism.[15] Based on a sample of one hundred borings, semispherical pits average an eighth of an inch (2.22 mm) in diameter and slightly more than a sixteenth of an inch (1 mm) in depth. There is no ready explanation for an inorganic origin of such pits with such consistent dimensions and a high surficial density. What kind of organism could make the borings in extremely hard rocks remains a mystery. The upshot is that additional investigation is needed on the potential of intertidal life during the Cambrian.

Another large Cambrian island along the Transcontinental Arch of Lochman-Balk covers present-day New Mexico, parts of western Texas, and western to southwestern Oklahoma (fig. 2.7). Defined by yet more unconformities, this region offers ample opportunity to explore smaller, related islands around the periphery. One such Upper Cambrian island is formed by rocks belonging to the Carlton rhyolite, which is chemically the same as granite but extruded as surface lavas. In the Slick Hills of southwest Oklahoma, Furongian deposits in the Honey Creek Formation consist of rhyolite rubble overlain by limestone composed of abundant echinoderm and trilobite fragments, followed by shale, including trilobites, brachiopods, and sponges.[16] Fossils show less wear and abrasion in more offshore deposits laid down in muddy water. Although fossil fragments are mixed among the rhyolite rubble at the island shore, no encrusting fossils are known to exist.

SUMMARY

The journey in Cambrian time westward from today's Massachusetts through Wisconsin and onward to South Dakota follows a line along a constant latitude of geography (fig. 2.1a), but the same passage five hundred million years ago stretched northward toward the Cambrian paleoequator crossing a smaller continent flooded by tropical seas. Taking a turn south at today's Black Hills in South Dakota, the journey ends in Oklahoma, but the same track during the Late Cambrian followed westward parallel to the paleoequator along a latitude roughly 10° south of the paleoequator. Sea level rose and fell during the later twenty to thirty million years of the Cambrian period, such that the width of the craton's Transcontinental Arch contracted or expanded with each fluctuation to drape a cover of eroded sediments higher and higher atop the Great Unconformity with Proterozoic bedrock.

Continental islands of all sizes appeared and receded under the ebb and flow of geologic time. Still, the Late Cambrian Baraboo Archipelago ranks as the premier example of an island group so exquisitely preserved that little imagination is needed to journey back in deep time and visit its shores. The varying shapes of smaller and larger islands on opposite sides of the archipelago resulted from an enormous structural bend in layers of underlying Baraboo quartzite that reached the surface as vertical limbs on one side and low ramps on the other. Prevailing trade winds from the southeast pushed waves toward the paleoequator and brought constant sea swells on a direct line of attack against one flank of the archipelago. Episodic hurricanes wreaked havoc from different directions, mostly from the northeast. Consequently, no part of the island group was spared from the slow wear of erosion against hard quartzite shores over time. Marine invertebrate life was plentiful beneath the low-tide line around the Baraboo monadnocks but was empty within the intertidal zone. The Cambrian islands of Laurentia are truly a primeval meeting place between the elements of earth and water as envisioned by Rachel Carson, but future research is sure to yield new details.

CHAPTER 3

HOW ISLANDS TRADE IN PHYSICAL WEAR AND ORGANIC GROWTH

A Journey in Late Ordovician Time to Hudson Bay's Jens Munk Archipelago

We think of rock as a symbol of durability, yet even the hardest rock shatters and wears away when attacked by rain, frost or surf.

—Rachel Carson, *The Edge of the Sea* (1955)

N o roads lead overland to Port Churchill on Hudson Bay in Canada's Manitoba Province. The town is serviced by train and air from Winnipeg in the south. Ship traffic reaches the port through the Hudson Strait from the east or the Northwest Passage from the west. With a permanent population of fewer than one thousand, the village exudes a frontier atmosphere composed of residents having Inuit, Cree, Chippewa, and European backgrounds. Inuit ancestors from the Thule culture roamed this part of the bay long ago, leaving a record dating back at least one thousand years. The northern limits of woodland Cree and Chippewa were established later on. Danish explorers hoping to discover a trade route to Asia overwintered on the opposite bank of the Churchill River from the future town site in 1619–1620 (see historical vignette, this chapter). The bastion of Fort Prince of Wales was built to last in stone overlooking Eskimo Island in 1731, marking a resolve for permanence by the Hudson Bay Trading Company (HBC). My link to Churchill derives from ancestors in the Sutherland family who immigrated from Scotland in the mid-1800s to work for the HBC. Fascinating as it became, a search for family roots was not what attracted me to Churchill in the summer of 1984. Rather, my initial visit hinged on a last-minute decision to repurpose the few remaining days of fieldwork in central Manitoba for an inexpensive train ride to the shores of

Hudson Bay. Not much was published on the local geology, but Ordovician and Silurian strata were said to be exposed along the coast (fig. 3.1).

After the port authority ceased operations for trans-shipment of wheat in 2016, Churchill's economy became fully refocused on ecotourism. Besides the historical attraction related to the HBC, wildlife excursions cater to summer visitors wishing to see pods of Beluga whales (*Delphinapterus leucas*) at the mouth of the

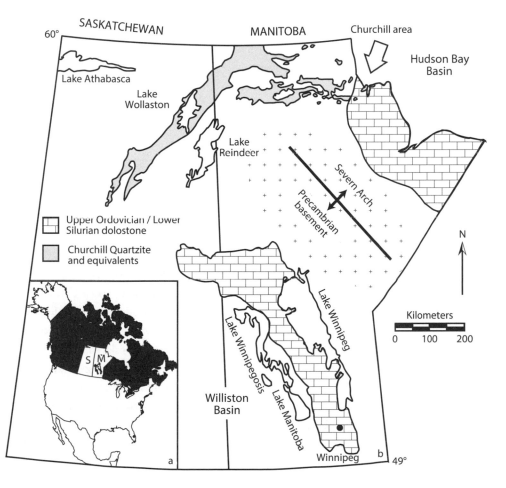

FIGURE 3.1 Ordovician-Silurian dolostone on basement rocks in central Canada: a) North America, showing the Canadian provinces of Manitoba and Saskatchewan; b) detail showing parallel exposures of dolostone on opposite sides of the Severn Arch, as well as location of the Paleoproterozoic Churchill quartzite (and equivalents). Large arrow points to Churchill, where Upper Ordovician strata surround quartzite monadnocks. Modified from Nelson and Johnson (2002, their fig. 1).

Churchill River and polar bears (*Ursus maritimus*) out on the tundra. Eating my lunch and staring out over the bay while seated atop a bluff at Bird Cove east of town, an unforgettable wildlife encounter occurred when a polar bear swam close below me. The animal cruised through the water with powerful strokes of its huge forepaws. Hikers are advised to carry air horns (like those on tractor-trailer trucks) that emit a loud noise capable of frightening off almost any living thing. On some summer nights, colorful displays of the pulsating aurora borealis may be witnessed. It all contributes to a place of natural drama and magic.

The most exquisite thrill I have known as a geologist struck me instantaneously at a level beyond anything else in the marvelous natural setting of Churchill on Hudson Bay. It came unexpectedly when, on the first day at midmorning, our small group arrived at the top of a quartzite sea cliff. The view over a becalmed sea was clear, and the tide was low. Small icebergs floated in the bay, holdovers from the deep freeze of winter. I saw a modern-day rocky coastline perfectly aligned with a preexisting rocky shore of immense antiquity. Had we arrived later during high tide, the extraordinary scene spread out below would have been submerged beneath the water and out of sight. Seaward of the massive cliff line, boulders of rounded quartzite sat immobile—encased in dolostone (altered limestone enriched in magnesium) and firmly fixed in place. I knew what it meant and was excited beyond words, eager to explore the bare rock surface on the open tidal flat. The cliff edge we stood on rose no more than 33 feet (10 m) above mean sea level. Overall, the prominence of the Churchill monadnocks on Hudson Bay is not as impressive as the quartzite cliffs around South Korea's Red Island on the Yellow Sea. The tallest monadnock in the entire region rises above the west bank of the Churchill River no less than 140 feet (~42 m). Also, the size of the Churchill monadnocks is rather small, with the largest covering less than 4,000 acres (16 km²). However, they are quite old as continental paleoislands. No more than 445 million years in age, the Churchill paleoislands lend witness to a time of transition near the close of the Ordovician period. Moreover, the story told by Churchill fossils is rich in ecological details representing marine life from both intertidal and shallow subtidal settings.

FEET ON THE GROUND

Resetting our geological clock for an advance in time by forty-five million years since departure from the Cambrian of Wisconsin's Baraboo Archipelago, we step ashore on yet another quartzite island caught in ecologic time near the end of the Ordovician period 444 million years ago (see fig. 1.9). A lot changed over those passing years. A substantial diversification known as the Ordovician radiation boosted the variety of marine invertebrates, as diversity doubled on an ordinal level and increased from about 700 known genera to 1,800 genera.[1] Less than a million years into the future, diversity would plummet during the first of five major mass extinctions to strike the blue planet. Far from Laurentia, conditions were building over lands centered on the south pole for a major glaciation that would put enough water into cold storage on land to drop global sea level enough to expose shelf margins elsewhere around that world. It is widely held that the reduction in living space enjoyed by marine organisms led to the catastrophic effects of the terminal Ordovician mass extinction.

The combined topographic and geologic maps for the Churchill district (fig. 3.2) provide the essential background for the area around the Churchill

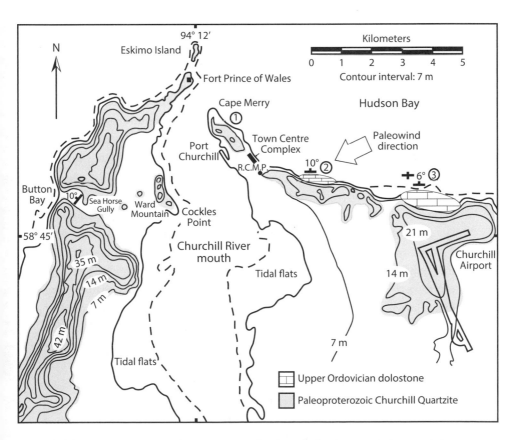

FIGURE 3.2 Distribution of quartzite monadnocks and related Upper Ordovician dolostone in the Churchill area at the mouth of the Churchill River, centered around Port Churchill (R.C.M.P identifies the Royal Canadian Mounted Police headquarters); Modified from Nelson and Johnson (2002, their fig. 2).

River estuary. At its entrance from Hudson Bay, the river mouth is only 1,000 yards (1 km) across. The narrow opening is reinforced by embankments of Churchill quartzite. Like the Baraboo and Sioux metamorphic rocks south of the Canadian border in Wisconsin and North Dakota, the Paleoproterozoic Churchill quartzite is derived from sandstone deposited 1.7 billion years ago, altered afterward to more durable quartzite. There is much to see close to town. Along Selkirk Street past the Anglican Church to the north, a left turn onto La Verendrye Avenue takes us to the front of the Town Centre Complex—which includes a hospital, public library, daycare facility, swimming pool, basketball▸ court, cinema, and cafeteria all under one roof. Large plate-glass windows on the north side of the building frame the view out over the open waters of Hudson Bay, less than 500 feet (~150 m) away.

On a pleasant July day in Churchill, the temperature can be expected to reach the average mark of 64.4°F (17°C), and there is better than a 50 percent chance of sunny weather. Under a light onshore breeze at midmorning, swarms of bothersome mosquitos are driven off the Churchill coast. Cape Merry is an easy walk to the west, 1.4 miles (2 km) from the Town Centre Complex (locality 1, fig. 3.2). The cape is protected by a line of low hills composed of gray Churchill quartzite that rises 46 feet (14 m) above mean sea level. Vernal pools trace fractures in the quartzite trending generally east–west. Looking across the channel toward Fort Prince of Wales, one can almost hear echoes floating on the breeze from the earlier commerce in the fur trade. There is no dolostone on Cape Merry, and the only such pocket anywhere on the opposite side of the Churchill River is found within Sea Horse Gulley off Button Bay (fig. 3.2).

Consulting a tide table back in the comfort of the Town Centre Complex, we are advised that the day's low tide is expected during the early afternoon. The maximum tidal range for this part of Hudson Bay is 15.5 feet (4.75 m). The farther the tide recedes, the greater the exposure of dolostone on the tide flats east of town. We pack no food for this hike. After an early lunch at the Town Centre café, we prepare for the rest of the day, carrying only water, cameras, and extra clothing in case of a sudden change in weather. Striking from the Town Centre Complex, a prominent quartzite ridge follows close against the shore to offer a clear view of the bay. The ridge climbs gradually in elevation for the first half mile (0.8 km), reaching a height commensurate with the knolls at Cape Merry. Deep fractures in the quartzite are exposed at the surface. On average, they intersect in two dominant directions: N 50°W by N 75°E. These trends make a noteworthy statement on how the coast is vulnerable to the erosion of storm waves, now and in the past. Moreover, overriding ice sheets during the last glaciations surely carried off pieces of quartzite from the basement rocks loosened along these lineaments.

Eastward, the first sign of a change is foretold by the midday sun as it shines on dolostone to reflect a dull beige tone in contrast to the dark quartzite. A line of flotsam is found after descending to the shore, tracing the high mark from the last big tide. Now, the tide is much receded to expose an ancient foreshore and the lime-rich layers that signify a former seabed (locality 2, fig. 3.2). As if by magic, the chilly waters of Hudson Bay are cast off to reveal a tropical shore over which we may walk at will. The object of immediate attention is a sea stack (fig. 3.3) that juts outward from its parent quartzite by 33 feet (10 m). Dolostone layers (formerly limestone) on opposite sides of the quartzite block register a 10° dip to the north. Like a splayed deck of cards, succeeding layers of limestone edge outward into the Hudson basin. In today's world, the calcium carbonate ($CaCO_3$) in limestone originates from organisms that typically thrive in warm waters between the equator and latitudes seldom more than 30° north or south of the equator. There are exceptions, but limestone-forming corals are limited by water temperature and the quality of sunlight that penetrates at high angles beneath the water's surface.

Often exceeding 6.5 feet (2 m) in diameter, the outermost rank of boulders marks the limit of strong wave energy against the former coastline. Here and

FIGURE 3.3 Upper Ordovician rocky shore on Paleoproterozoic Churchill quartzite. Sea stack in foreground is 13 ft (4 m) wide and 33 ft (10 m) long, surrounded by dolostone.

Photo by author.

there, the big boulders are identified by numbers painted during a mapping survey conducted in 1985.[2] Boulder number 11 is spotted about a dozen feet (~4 m) east of the massive sea stack. All such boulders are firmly cemented in dolostone, but they rise above the surface by as much as 3 feet (~1 m) or more. The scenario is wonderfully three-dimensional in effect. Time travel back to another world could hardly be more realistic. A short space landward, the density of quartzite clasts embedded in dolostone exhibits a chaotic mix of cobbles and boulders that form a substantial conglomerate (fig. 3.4). Normal sea swell reaching the coast would hardly possess sufficient energy to move the boulders, but smaller cobbles and pebbles would be in constant motion. Likewise, the largest stones were eroded from parent quartzite and jostled into place by major storms. Loose cobbles have a distinctive heft, with a measurable density of 2.7 times that of the standard for water (1 gm / cm^3). Pausing here, one can imagine the intensity of swirling eddies and the noisy clatter made by dense quartzite clasts of all sizes colliding against one another. Angular stones lose their sharp edges to become

FIGURE 3.4 Upper Ordovician conglomerate derived from the Churchill quartzite within a dolostone matrix (hammer for scale).

Photo by author.

smooth through this process. Subtle details pull us back to a harsh northern shore where the shattering effect of frost and the impact of sea ice left a gaping hole in an enormous quartzite boulder (fig. 3.4).

Few painted numbers survive after so many years subject to the diurnal wash of the northern tides, but thirty-nine boulders were so identified in an undulating line that extended for 1,150 feet (350 m) adjacent to the modern sea cliffs. Part of the original survey is reproduced (fig. 3.5) to show the location of boulders 18 to 31 in the progression. Some are very large. Numbers 18 and 22 sit offshore a prominent headland in the parent quartzite bluff, each exceeding 26 feet (8 m) in diameter. The numbers may have vanished, but the outsized boulders remain solidly in place as markers along the former shoreface. Wandering steadily eastward but back and forth in a zig-zag fashion with the boulders as a constant reference

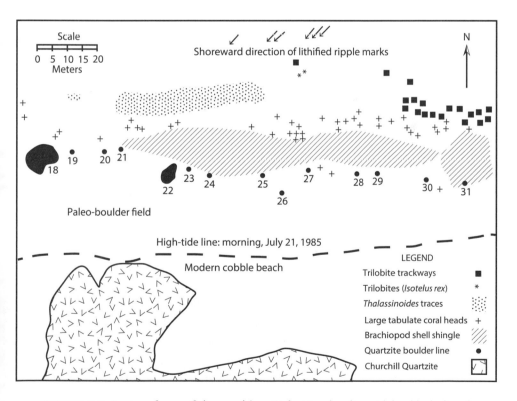

FIGURE 3.5 Portion of a 1,150-ft (~350-m) long Ordovician shoreline with boulder beds and limestone strata exposed on the modern tidal flat during extreme low tide. Black dots mark seaward location of quartzite boulders with minimum diameter of 6.5 ft (2m) locked in dolostone. Modified from zonal scheme of Johnson et al. (1988, their fig. 7).

point, it is striking to see how the volume of eroded quartzite is reduced in the seaward direction. Roughly 33 feet (10 m) offshore, the average size of quartzite fragments encased in dolostone layers is that of a pebble (3/8 in or 8 mm in diameter). At a further distance of about 65 feet (20 m) seaward of the line, quartz sand is visible only with a hand lens.

A vexing complication in time and scale must be confronted because the map (fig. 3.5) plots details from different layers that dip evenly offshore, each representing a slightly younger timeframe as one works upward through approximately 15 feet (4.5 m) of layered dolostone. That is, if the outermost layer accessible at maximum low tide were projected back toward the paleoshore through its dip angle, that layer would arch into the air well above the line of big boulders. Another way to explain it is that all the dolostone beds seaward of the reference line are beveled by erosion. The effect foreshortens the actual distance between the various features from different time frames plotted north to south on the map. Hence, the actual gradient in the reduction of clast size as a function of distance from the parent quartzite shore is greater than implied by the map. The phenomenon is familiar to geologists as Walther's law, which posits that different facies (physical and/or fossil aspects of rock layers) shown to be lateral to one another in horizontal space also follow above one another through time. In brief, the conglomerate and former limestone layers at this locality on Hudson Bay represent a relative rise in sea level over a short interval of geologic time. The whole scenario is contrary to the major drop in sea level, yet to come, which brings the Ordovician period to a conclusion.

Changes in the amount and coarseness of eroded quartzite from layer to layer through any outward traverse across the tidal flat are impressive. So too are changes in the macrofossils from one band of distinctive rock to the next as far as the receding tide allows. Among the most easily recognized are the large coral colonies. They are not so abundant and interlocked so as to form a reef structure, but they are hard to miss. Other features that count as fossils are the trails and burrows left by marine animals that lived on or within the seabed. Among these are the extensive horizontal networks of tunnels belonging to the trace fossil *Thalassinoides*. The maker left no fossilized body parts, but it is assumed to have been a species that behaved something like today's ghost shrimp that commonly constructs similar burrows.

No designation is agreed to by geologists regarding a particular name for the related conglomerate and dolostone layers so well exposed at this place. In the same sense that the Tunnel City Group defines the conglomerate and sandstone beds from the Baraboo district in Wisconsin, the sedimentary rocks from this part of Hudson Bay are called the Churchill River Group. As in Wisconsin, the difficulty arises in correlating such locally distinctive rock formations from around island monadnocks with other more widespread rock formations in outlying districts. Even so, the stratigraphic position of the conglomerate and limestone beds peculiar to the Churchill district is agreed by most geologists to correlate with strata in the Upper Ordovician. Microfossils used for correlation include certain conodonts (*Oulodus rohneri* and *Pristognathus bighornensis*) recovered from samples at this spot. Conodonts are the distinctive phosphatic elements of mouth parts belonging to small, free-swimming animals closely related to primitive fish lacking true jaws. Proficient as swimmers, the conodont animal left excellent index fossils.[3]

In part, the Churchill River Group of Manitoba is correlated with the Hirnantian stage, which occupies the uppermost part of the global Ordovician system. The corresponding Hirnantian age is an interval of geological time that lasted about two million years until the close of the Ordovician period 443.7 million years ago (see fig. 1.9). Under the authority of the International Union of Geological Sciences, all igneous rocks and sedimentary strata formed during that time anywhere in the world are regarded as belonging to the Hirnantian stage. The name derives from a place in Wales called Cwm Hirnant, which translates as "valley of the long stream," where the Hirnant limestone appears at the top of Ordovician strata. The name is conserved, but the reference locality for the Hirnantian stage is now redefined as a locality in China on the basis of graptolite fossils.[4] Such fossils are unknown in the Churchill district, and the Hirnantian reference locality in China includes no diagnostic conodonts. The matter is complicated because other fossils more characteristic of strata from the overlying Silurian System also appear in the Churchill River Group. However, correlative strata in China record carbon isotopes showing a major positive excursion, since recognized in many other places worldwide, including the Churchill district.[5]

Additional visits to the bay are worthwhile but must be carefully timed with the tidal cycle. Details regarding the most unusual fossils from the conglomerate and dolostone layers merit further discussion (as follows), but the afternoon is late, and our remaining time is limited by the rising tide. A return route south across the adjoining ridge of Churchill quartzite (fig. 3.2) affords the opportunity to gauge the monadnock's midsection. In a direct march, the distance is eight-tenths of a mile (~750 m), taking us across the highest part of the ridge with an elevation slightly more than 70 feet (~21 m) above mean sea level (higher than Cape Merry). It's a surprise, perhaps, that no conglomerate or dolostone beds from the Churchill Group are exposed along the monadnock's south face. The road west back to town intersects with La Verendrye Avenue past the headquarters for the Royal Canadian Mounted Police.

INTERTIDAL AND SUBTIDAL CORALS

The boulder field caught in ecologic time by the Upper Ordovician conglomerate invites hours of fossil hunting in search of corals cemented on quartzite in their original growth position. During the 1985 field season, the prediction was made that boulder-encrusting corals should be found here. The seed for my intuition was planted months earlier the same year during a visit to the British Virgin Islands, where living corals were observed in shallow, aquamarine waters attached to pink granite. Scarcely within ten minutes of my suggestion, a discovery was confirmed for the first such fossil coral from Churchill.[6] In cross section (fig. 3.6), the actual colony from the initial find shows more than fifty septa (where individual coral polyps were seated), each with a ladder-like subdivision (called tabulae) that progressively sealed off the floor of the calcified cup (calyx) in which an individual member of the colony grew. Identification to species level would not be possible without collecting the fossil to make a petrographic thin section, but the resemblance to the common honeycomb coral (*Favosites* sp.) is unmistakable. The coral grew in place for perhaps a dozen years in a low and disc-shaped colony with a diameter of 6 inches (15 cm).

Once the first example was discovered, others quickly followed. It became a challenge to locate coral colonies attached to a substrate having more relief

FIGURE 3.6 Cross-section through a tabulate coral (*Favosites* relative) 4 in (10 cm) across on a flat quartzite surface.

Photo by author.

than a flat surface. Indeed, some corals also grew on quartzite knobs, giving an appearance of a colony more club-like in shape.[7] Others showed where a colony was partially eroded from its attachment site to reveal a carbonate residue with small-diameter borings that penetrated beneath the remaining coral. Living coral colonies are commonly bored by various organisms, including sponges, marine worms, and bivalves. In this case the Churchill corals likely attracted dwellers from among the marine worms (tube-dwelling Polychaeta). Later on, another type of coral occupying the same rocky shoreline was verified in addition to the honeycomb variety. Evidence based on thin sections conclusively shows that a heliolitid tabulate coral (*Ellisites* sp.) grew in thin sheets on quartzite.[8] Original surfaces on eroded quartzite boulders also exhibit circular depressions up to an inch (2.5 cm) in depth, often filled with Ordovician dolostone.[9] Jellyfish are related to sea anemones, and fossil jellyfish similar to those from the Cambrian of Wisconsin are known from Hirnantian strata elsewhere in Manitoba.[10] Sea anemones probably occupied a niche in the Late Ordovician intertidal zone on the quartzite islands of the Churchill district.

In contrast to the boulder dwellers, other corals thrived offshore in a setting that was deeper and arguably less agitated by wave surge.[11] The zone features colonies as big as 18 inches (~46 cm) in diameter that were freestanding and unattached to the seabed. A branching coral (*Palaeophyllum* sp.) is one such example (fig. 3.7), which, in this case, was preserved intact but pushed over on its side. Generally, the larger size of such corals prevented them from being moved around on the seabed by all but the largest storm waves. Another coral morphology typical from this zone is represented by heavier, dome-shaped colonies from at least two kinds (*Calapoecia* sp. and *Catenipora* sp.). Isolated quartzite clasts up to 3 inches (5 cm) in diameter are scattered within the same layers among the large corals. During low tide, the big corals can be traced in a zone along the full 1,150-foot (350-m) expanse of dolostone exposed west to east on the tidal flats.

COASTAL MARINE ZONES

Both landward and seaward of the freestanding corals, other zones are well exposed on the same tidal flats (location 2, fig. 3.2). Shoreward but outside the line with large quartzite boulders (fig. 3.5), bedding surfaces are shingled by large shells belonging to a brachiopod (*Monomerella* sp.). Unlike most clams (bivalve mollusks), living brachiopods possess a fleshy stalk that attaches the

FIGURE 3.7 Cross-section through free-standing tabulate coral (*Palaeophyllum* sp.) approximately 18 in (~ 46 cm) across.

Photo by author.

animal to the seabed. In this case, the fossils are preserved as molds that show the impression of a unique muscle scar on one of the valves. As previously mentioned, the zone with extensive trace fossils (*Thalassinoides*) located seaward from the freestanding corals belongs to a burrow-maker similar to today's ghost shrimp.

Seaward beyond the fossil burrows is a zone that includes trilobites and their trackways. The world's largest trilobite (*Isotelus rex*) was collected from this locality, measuring 27.5 inches (70 cm) in length.[12] The trilobite I stumbled upon during our 1985 field season measures almost 17 inches (43 cm) from head to tail (fig. 3.8a). The larger specimen now resides in the collections of the Manitoba Museum in Winnipeg. The smaller specimen is held in the National Type Fossil Collection at the Geological Survey of Canada in Ottawa. These fossils are not only national treasures but world treasures. Because ice floes loosen and pry

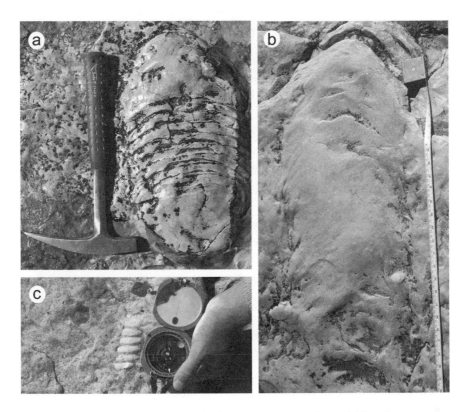

FIGURE 3.8 Megafossils from Upper Ordovician strata exposed on tidal flats during extreme low tide: a) Large trilobite *Isotelus rex* (hammer for scale); b) trilobite trackway showing traces left by the forward shuffle of cephalic shield and blunt genal spines. Tape measure for scale extended 40 in (1 m); c) siphuncular fragment of a large orthocone cephalopod (compass case for scale, also recording orientation).

Photos by author.

apart layers of dolostone every winter, more of these giant trilobites are sure to be found. A related discovery entails a fossil trackway (fig. 3.8b) that corresponds to the body width of the big trilobite. The trace is special not only because of its length (40 in or 1 m) but because it shows the arc of the trilobite's huge head shield and markings after the pair of matching spines at the side of the head. The animal's movement across the seabed is construed to have taken place as a shuffle that clearly recorded its halting movements. Where it ends abruptly, the track suggests that the trilobite flexed its middle section (thorax) and used its head shield to tunnel beneath the sandy-carbonate surface.

A REGISTRY OF WINDS

Trilobites from the phylum Arthropoda were not the only large marine invertebrates to populate the coastal waters off the Churchill monadnocks during the Late Ordovician. An important group within phylum Mollusca is the class Cephalopoda, which includes today's octopus, squids, and the chambered nautilus. The nautilus shell is planispiral in design with an interior that reveals a series of compartments connected by a partly calcified organ called the siphuncle. Biology tells us this sort of cephalopod uses its siphuncle to regulate buoyancy through the flow of dense liquid to and from the chambers. Thus, the animal functions like a kind of submarine capable of moving up or down through the water column. Cephalopods also have a tube-like structure (hyponome) located below the head, through which water may be expelled under pressure for jet-like backward swimming. Ordovician nautiloids possessed a chambered shell, but it was long and cone-shaped. The maximum size for an Ordovician orthocone is about 5 feet (1.5 m). In some Ordovician nautiloids, the siphuncle was unusually large and well calcified with bulbous segmentations. The enclosing shell with its multiple chambers was more fragile and subject to postmortem disintegration leaving behind the remains of the heavy siphuncle.

Fossil nautiloids in the Churchill River Group typically occur as siphuncular fragments that look superficially like coarsely ribbed broomsticks of varying length (fig. 3.8c). In a survey by Skinner and Johnson (1987), the locations and orientations of 189 such fragments were plotted over the entire tidal flat at locality 1 (fig. 3.2).[13] Results showed that the siphuncular segments are closely aligned with a preferred orientation perpendicular to the paleoshore. Combined with available data on the orientation of ripple marks preserved on a surface in the outer part of the tidal flat (fig. 3.5), it is argued that wind-driven waves arrived from an oblique angle to the coastline and rolled the heavy siphuncular fragments along the shore in a consistent direction from present east to west. The waves impacting this part of the archipelago produced both a high-energy boulder zone and an effective long-shore current.

OTHER ISLANDS IN THE JENS MUNK ARCHIPELAGO

Another embayment with extensive dolostone is situated north of the Churchill airport (locality 3, Fig. 3.2). The airport occupies a quartzite monadnock that covers 2,600 acres (~10.5 km²). Most of the 11,500-foot (~3.5-km) concrete runway sits 80 feet (~24 m) above mean sea level. The tidal flats at Airport Cove are immense, with layers covering as much as 250 acres (~1 km²), fully exposed only during extreme low tide. The conglomerate zone so well developed in the adjacent bay is absent. Overall, the dolostone layers at Airport Cove are inclined at a lesser tilt. Close to shore, the beds dip 6° to the north but quickly level off to flat-lying beds.[14] Owing to the flatness of the layers, the distinctive zonation of coastal zones so well developed in the neighboring bay is not evident. A loose block of limestone filled with the articulated shells of a large brachiopod (*Virgiana decusata*) preserved in growth position was recovered from this locality, which implies that some strata may range upward into the overlying Silurian system. The most remarkable fossil documented from Airport Bay is the oldest horseshoe crab (*Lunataspis aurora*) on record.[15] Fragmentary pieces of sea scorpions (eurypterids) also are reported here. These significant discoveries expand the source of trackways preserved in strata from the Churchill River Group to additional animals. Other unusual fossils from Airport Cove include lightly calcified green algae (dasycladacean algae).

As many as four island monadnocks of varied size occur farther to the east along the coast. From a geographic perspective, it is most telling that strata are inclined outward from opposite sides of a particular monadnock. In this case, it is as close as we may come to finding a quartzite paleoisland surrounded by the former limestone. Strata with abundant brachiopods (again, *Virgiana decusata*) are preserved in a growth position in this area.[16]

HISTORICAL VIGNETTE: JENS MUNK AND THE DISASTROUS VOYAGE OF 1619–1620

The average daytime temperature for Churchill in July is not much different than for Copenhagen, Denmark. But the local climate is unstable, and the record July low at Churchill is 28°F (−2.2°C). Our team experienced sleet on a July afternoon out on the tidal flats. Denmark enjoys maritime conditions, which are milder in winter than Canada's continental climate. The average January high for Churchill is 22°F (−7.4°C), whereas Copenhagen experiences an average January high of 32°F (0°C). The record January low for Churchill is minus 45.6°F (−50°C), a mark unheard of in Denmark. Such is the background for the disastrous voyage of 1619–1620 in search of the northwest passage to Asia commanded by Captain Jens Munk on behalf of Denmark's King Christian IV. The expedition was provisioned with supplies intended to last through a two-year voyage. From the start, it was planned to establish a winter camp somewhere on Hudson Bay, which was mistakenly thought to enjoy a climate not unlike Denmark. The captain's journal tells the story of an epic disaster in the bid to find and exploit a new trading route with Asia.[17]

On May 25, 1619, the expedition left Copenhagen with a company of sixty-five men in two Danish warships, the larger frigate *Unicorn* and the smaller sloop *Lamprey*. Only three men, including Captain Munk, would survive to reach civilization in Bergen, Norway, on September 27, 1620. Rations of beer and spirits never became a problem during the expedition, and sufficient gunpowder and arms were brought along to support hunting parties during the winter. The *Unicorn* reached a safe harbor near today's Cockles Point on the west bank of the Churchill River (fig. 3.2) on September 7, 1619, and the *Lamprey* arrived the following day. Some members of the crew were already ill. Munk ordered a landing party to collect gooseberries, cloudberries, and *tydebaer* (small, red berries much like cranberries), which he knew from childhood in Norway. All such fruits are rich in vitamin C, the essential dietary deterrent against scurvy. That same day, whale meat was acquired from a "beluga fish," but the next morning, the crew was confronted by a polar bear intent on taking the meat. Munk shot the bear and had its meat boiled and treated with vinegar. "It was of good taste and quite agreeable," he wrote. Early on, hunting parties brought back rabbits and ptarmigans daily.

November 7, 1619, Munk set out with nineteen men to explore inland but was forced back by a snowstorm. He realized it would be impossible to travel any distance from the ships during the winter without skis. The first sailor died on November 21, 1619, and afterward, during the new year, deaths occurred daily. There was an insufficient larder of berries to protect such a large group against the ravages of scurvy. The place Munk named Nova Dania would be their home until the breakup of ice on the river and departure on July 16, 1620. The *Lamprey* was refloated, and in an astounding feat of seamanship, it was sailed by the three survivors through the treacherous Hudson Strait and back across the North Atlantic to Norway. The financial loss to King Christian IV was substantial, and Munk was left in Bergen for a year before being summoned back to Copenhagen. The ancient monadnock islands of the Churchill district were named the Jens Munk Archipelago in honor of an intrepid explorer who faced and overcame countless obstacles during an eventful life.[18]

ISLANDS ELSEWHERE IN THE ORDOVICIAN WORLD

Like the Churchill quartzite, the Lorain quartzite from Manitoulin Island in the Great Lakes district of Canada's Ontario Province is yet another 1.7-billion-year-old sand body affected by metamorphism. These durable rocks shaped regional topography for millions of years through the later Ordovician period. The Sheguiandah area in eastern Manitoulin Island is another place where the relationship between island monadnocks and Ordovician strata is well exposed. Detailed work on the basal Collingwood Formation yielded a distinctive fauna of brachiopods (including species of *Dalmanella, Fardenia, Paromalomena,* and *Triplesia,* among others) and a trilobite (*Triarthrus canadensis*) correlated with the Upper Ordovician.[19] This deposit is not quite Hirnantian in age but very close. Compared to Churchill, the trilobites are tiny and rarely complete. The brachiopods, too, are small but well preserved, with both valves intact. A complete growth series for the most abundant brachiopod (*Dalmanella* sp.) ranges through individuals between 1/16 and 3/8 inch (1.3 and 12 mm) in shell size. Known from only a single locality in the Sheguiandah area, the fauna was buried against the side of a plunging rocky shore on an island monadnock formed by Loraine quartzite. In contrast to the intertidal coral fauna from the Churchill district, the trilobite and diverse brachiopod fauna lived in deeper water well below the zone of agitation by surface waves.

The Cliefden Caves area of New South Wales features Upper Ordovician strata consisting of limestone breccia from a carbonate platform that encircled a chain of volcanic islands on the Molong High.[20] The age of the paleoislands is imprecise, but like the Collingwood Formation of Ontario, they occur at the top of the Ordovician system, not far off the Hirnantian stage. The underlying basement rock at the core of the Australian paleoislands is andesite. From available studies, it is unclear if the breccia was related to volcanic or storm activity.

ORDOVICIAN GLOBAL GEOGRAPHY

During the Ordovician period, the ancestral North American continent of Laurentia began a clockwise rotation from 90° off the continent's present orientation to one around 45°. Against this backdrop, the Churchill district was located 10° south of the Late Ordovician equator and was subject to strong trade winds that emanated from the southeast to drive surf against the windward flanks of the Churchill monadnocks (fig. 3.9). Tropical storms impacted the Churchill district but perhaps not with the same intensity as the hurricanes that struck the Baraboo Archipelago regularly during the Late Cambrian period. Hurricanes were more likely to have reached the Great Lakes district of present-day Ontario, with its monadnocks clustered at a latitude of 20° of south of the paleoequator.

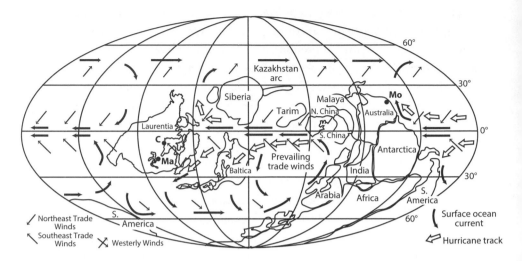

FIGURE 3.9 Global reconstruction for the Late Ordovician (Hirnantian) world with interpretations for trade winds convergent on the paleoequator, high-latitude westerly winds, surface
ocean circulation, and predicted hurricane tracks. Abbreviations: C = Churchill; Ma = Manitoulin
Island; Mo = Molong High.

At a larger scale of paleogeography, all of Australia, including present-day New
South Wales, was formerly a part of the megacontinent Gondwana. Antarctica,
India, Arabia, Africa, and South America comprised the rest of that continent, part
of which was located over the Ordovician South Pole. During Hirnantian time,
major ice caps advanced from the South Pole, affecting sea levels everywhere. The
volcanic island chain along the Molong High sat on a continental shelf exposed to
the open ocean at a latitude more than 15° north of the paleoequator. Late Ordovician islands in the present-day Cliefden Caves area would have been subject to
strong trade winds that emanated from the northeast, as well as hurricanes with a
propensity to strike the continent from the southeast (fig. 3.9). Wind-driven waves,
whether due to constant trade-wind circulation or more episodic hurricanes,
provided the raw energy to chip away at the most resistant rocky shores.

SUMMARY

Countless islands, big and small, populated the Ordovician world, just as they
do today. Metamorphic rocks like quartzite provided the resistant basement
rocks that formed the core of islands throughout the Jens Munk Archipelago

in present-day Manitoba and similar islands in Ontario. Geologically, the break between older quartzite and sedimentary strata that trace rocky shores around island monadnocks gives us a solid sense of geography hard to match elsewhere in the Ordovician world. The Upper Ordovician relics of the Jens Munk Archipelago are sublime for their precise physical lines and for the rich fossil heritage that they yield regarding intertidal life as well as offshore biological zones regulated by variations in water depth and turbulence. Not least of all, the Churchill scenario is extraordinary for the way it fits global Ordovician geography to mesh with patterns of ancient ocean currents, prevailing trade winds, storms, and changing sea levels.

It was a privilege to work with Canadian field geologists like Sam Nelson and to bring students and other colleagues to Churchill to explore the Jens Munk Archipelago. Among experts on Upper Ordovician strata, Professors Rong Jia-yu (Nanjing Institute of Geology and Palaeontology) and Stig Bergström (Ohio State University) made the trip during the summer of 1987. The challenge arose to see who among us might last the longest after a plunge from the Ordovician shore into the icy July waters of Hudson Bay. The Scandinavian stayed in the longest. Professor Rong left with the distinction of being the only Chinese in the long history of his culture to have reached these shores and endured these waters. On his return to China, Professor Rong informed me about exposures of Silurian rocky shores from Inner Mongolia that I truly needed to see for myself.

Following the onboard tracking system during a midwinter flight from the United States to Asia, my next return to Churchill was a flyover well in sight of the coastline. Hudson Bay was frozen over in solid ice. How ironic, I thought, that Churchill polar bears range out on the ice from dolostone tidal flats that represent former tropical shores. The dense Churchill quartzite is not immune to the slow ravages of time that debase the shoreline now as then but has persisted as the foundation of well-defined paleoislands lasting 445 million years. Those ancient isles preserve an astonishing array of fossils characteristic of the great Ordovician radiation. How peculiar, I thought, that the giant trilobite (*Isotelus rex*) would go extinct, but the ancestor of the horseshoe crab (*Lunataspis aurora*) would spawn surviving descendants well known by Rachel Carson on her perambulations along Carolinian sandy shores during low tide.

CHAPTER 4

HOW ISLANDS RECALL WINDWARD SURF AND LEEWARD CALM

A Journey in Late Silurian Time to
Inner Mongolia's Bater Island

On all these shores, there are echoes of past and future: of the flow of time, obliterating, yet containing all that has gone before; of the sea's eternal rhythms—the tides, the beat of the surf, the pressing rivers of the currents.

—Rachel Carson, *The Edge of the Sea* (1955)

The evening sky had darkened by the time the 8:00 P.M. train left the central Beijing station bound for Hohhot, the capital of the Inner Mongolia Autonomous Region on China's frontier with the sovereign state of Mongolia—often called Outer Mongolia. There was little of note in the Beijing suburbs, but afterward, the tracks ran parallel to parts of the Great Wall of China. We were scheduled to arrive in Hohhot, the Green City, at 7:10 A.M. on Monday, July 5, 1999. Hohhot would be our gateway to the Joint Banner Darhan-Mumingan district, an area normally off-limits to Western visitors. Arrangements had been made in advance, and the necessary permits issued to allow access to a place where the populace had experienced few outsiders. Elsewhere, much of China had been open to tourists long before. Our principal host, Li Wenguo, an experienced field geologist employed by the Inner Mongolian Geological Survey, would meet us. In the coming days, our field party would swell to eight members plus the drivers assigned to convey us in two sturdy vehicles equipped for off-road travel. The vast area of the Autonomous Region is China's broadest province, stretching some 1,500 miles (~2,400 km) northeast to southwest, from forested mountains to central grasslands to desert sands. Li Wengou covered much of it on mapping expeditions with logistics supported by trucks, pack horses, or camels as appropriate in different terrains.

Days earlier, I was the guest of the Geology Department at Beijing University, where I made a presentation on rocky-shore ecosystems with images from the Upper Ordovician of Churchill, Canada, and other places in Mexico and Australia. The American academic Amadeus William Grabau was a professor of paleontology at the National University of Peking (as then known) from 1920 until 1941. During his tenure, the first generation of paleontologists in China received advanced training. Among his students was the mentor of my colleague and friend Rong Jiayu in Nanjing. Grabau's gravesite occupies a peaceful spot near a lotus pond on the campus of Beijing University. Our delegation paid respects, leaving a bouquet of flowers. Among other accomplishments, he introduced a set of paleogeographic maps that interpreted Ordovician and Silurian icecaps in the southern hemisphere.[1] Comprehensive for its time, he also compiled a sea-level curve based on data gathered from Paleozoic strata worldwide.[2] Now from Hohhot, we prepared to leave northward for the border area with Outer Mongolia in search of rocky shores that weathered the rise and fall of Silurian seas peculiar to the geography of a microplate called the North China Block.

This chapter examines the range of physical and biological relationships on and around a single, small continental island of Silurian age named Bater Island.[3] The name derives from a nearby landmark celebrated as Bater Obo, but the word *bater* in the local dialect also translates as "hero." Hence, Bater Island may be construed in English as Hero Island. In Mongolian culture, an obo is a place of reverence marked by a stone cairn or other structure, often at the top of a mountain. The sacred mountain and its related paleoisland are located east of the Xibiehe River, about 100 miles (~160 km) northwest of Hohhot (fig. 4.1a).

Leaving Hohhot, the paved road climbed a series of switchbacks to reach a pass through the hills with access beyond to mixed-use agricultural and pastoral lands. Ethnic Han Chinese account for nearly 80 percent of the provincial population, whereas Mongol peoples comprise 17 percent. Historically, there has been pressure to expand farmlands farther north, edging out traditional grazing lands. Han farmers plant maize (corn) in competition with Mongol shepherds, who mostly tend sheep. Darhan, the district's main town, is where we established team headquarters. On the town's northern outskirts, the slogan painted on a vertical slab of concrete read (in large Chinese characters): "Protect Grassland Resources." On the other side, an image of the globe showing much of Asia was accompanied by text (also in Chinese): "Human beings have only one Earth."

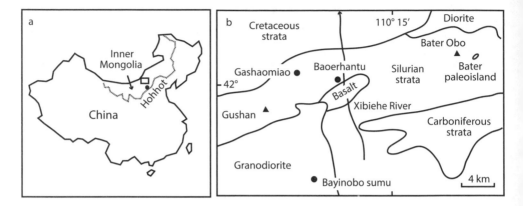

FIGURE 4.1 Maps covering part of the Joint Banner Darhan Mumingan district in northern China: a) relationship of the Inner Mongolia Autonomous Region to the rest of China with inset box denoting study area; b) detailed geographic and geologic map for the areas east and west of the Xibiehe River.

From that point, the uninterrupted grasslands rolled northward to the frontier with Outer Mongolia.

FEET ON THE GROUND

Centered around the village of Baoerhanto, the Joint Banner Dahrhan-Mumingan district covers an area exceeding 250 square miles (650 km²). Roughly a quarter of that area is taken up by exposures of igneous rocks, including granodiorite, basalt, and diorite (fig. 4.1b). Unlike basalt, diorite is more akin to granite as an igneous rock that cooled slowly within a buried magma chamber long before exposure at the surface by erosion. Diorite has a composition of coarse minerals that differs from granite with significantly less quartz silica. Silurian strata in this region are assigned to the Xibiehe Formation, named in acknowledgment of the Xibiehe River as a local landmark. These strata abut all three types of igneous rocks from place to place, effectively tracing the contours of former rocky shores. Near Baoerhantu, the contact between Silurian strata and basalt includes a sea stack surrounded by basaltic conglomerate. In contrast, Silurian strata farther west of the same village are dominated by sandstone that adjoins the same basalt but sourced from silica-rich granodiorite located farther to the west.[4]

Bater Island is a sedimentary sheath wrapped around a monolith of Ordovician diorite. The igneous rock gives the place its surviving topographic stature, exhumed from beneath a cover of younger Silurian strata still enclosing it on three sides. East of Baoerhantu (fig. 4.1b), the feature sits alone on the open grassland in a natural state without any hint of human presence. To reach it, we drove off-road across the rolling countryside. Seen to best advantage from the south (fig. 4.2a), the island's elliptical shape is distinct owing to the dark tone of naked igneous rocks in contrast to the lighter shade of surrounding limestone. On approach from the southeast, the grasslands conceal underlying rocks until the last moment. Preparing to set foot on the rocky shore, we set our geological clock to 420 million years ago. Over 25 million years have elapsed since we left the Late Ordovician monadnocks around Churchill on Canada's Hudson Bay (see fig. 1.9). Sea levels rebounded after the severe draw-down at the close of the Ordovician period, and marine invertebrates regained strength in numbers by an additional 150 genera.[5] Primitive land plants dominated by moss were in place by this time but restricted to galleries kept wet by stream water. The smallness of rocky Bater Island precludes such a distinction.

The geologic map for Bater Island lacks a superimposed topographic map (fig. 4.3a), but a cross-section through the paleoisland provides some perspective on physical relief (fig. 4.3b). Overall, the maximum length of the island is 2,000 feet (610 m) against a width of 655 feet (200 m). It is a tiny island compared to many in the older Jens Munk Archipelago of Manitoba (see chapter 3) and the Baraboo Archipelago of Wisconsin (see chapter 2). The exposed vertical relief amounts to merely 100 feet (~30 m), but the relief is likely to be much greater hidden below the grasslands.

Here, thin layers of siltstone pass into thicker beds of silty limestone that recline against the diorite with bedding that dips to the southeast at high angles between 40° to 50°. The climb on foot to the center of the island is steep at first (fig. 4.4a), where the sedimentary cover is thin.

Large fossil stromatoporoids, dome-shaped in construction (figs. 4.3a; 4.4b), sit in their original growth position on or very near the diorite surface. These calcified sponge-like colonies form a distinct band that may be followed laterally along the island's southeastern flank for about 330 feet (100 m). In places where the domes are detached, it is possible to see the basal scar revealing a growth pattern on the unconformity surface (fig. 4.2b). The surrounding dark siltstone is enriched in the minerals plagioclase, chlorite, and augite derived from direct erosion of the diorite. Elsewhere on or close to the unconformity surface, it is possible to find patches of broken crinoid stems (fig. 4.5), some as much as 3/16 inch (1 cm) in cross-section.

FIGURE 4.2 Perspective view with detail of the Bater Paleoisland: a) South flank of the complete diorite outlier (dark center) exhumed from overlying Silurian strata to expose the core of Bater paleoisland; b) close-up showing the attachment site with concentric growth pattern of a stromatoporoid colony sitting on the unconformity surface (camera lens cap for scale = 2.5 in or 7.5 cm in diameter).

Photos by author.

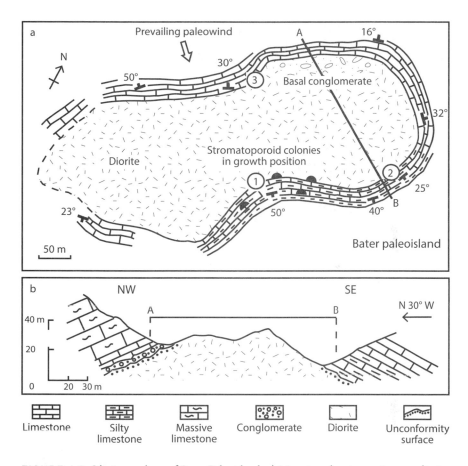

FIGURE 4.3 Silurian geology of Bater Paleoisland: a) Map view showing an igneous diorite core surrounded by flanking limestone dipping away from the core in all directions; b) Cross section through the east side of the paleoisland (transect A–B), showing relative topography and differences in abutting rock formations.

Exploring the northeast end of the structure (fig. 4.3a), the silty limestone characteristic of the southern slope thins to merge with bedded limestone lacking any silt content. Here, sedimentary layers above the diorite reside on a lesser slope, dipping away between 25° to 32°. A minimum dip angle of 16° is encountered in the far north corner of the structure. Continuing a counterclockwise tour around the paleoisland, a distinct change in lithology appears as a basal conglomerate formed by eroded cobbles of diorite that extends in a band for nearly 500 feet (150 m). Small boulders within the conglomerate measure as much as 15 inches by 9 inches (38 cm by 23 cm) in diameter. Farther along,

FIGURE 4.4 Southeast flank of Bater paleoisland: a) geologists (Johnson, Baarli, and Li from left to right) stand on the sloping unconformity between diorite basement rock and overlying silty limestone (see previous figure 4.2b); b) Cluster of large stromatoporoid colonies (*Plexodictyon* sp.) from 6 to 12 in (15 to 30 cm) in diameter in growth position on the unconformity surface.

Photos by Wang Jian (a) and by author (b).

FIGURE 4.5 Scattered crinoid columns in basal Silurian strata sitting on the unconformity with basement diorite at the core of the Bater paleoisland. Unlike other fossils preserved in growth position, the crinoid stems represent debris left in place.

Photo by author.

the conglomerate disappears, and the limestone above the contact takes on dip angles between 30° to 50° slopping away to the northwest. Closing the circuit, limestone layers that flank the south end of the structure register a dip angle of 23° to the south (fig. 4.3a). No faults with obvious displacement of basement rocks and adjoining strata are encountered anywhere along the structure's perimeter. At its narrowest, the top of the structure reveals barren diorite exposed for 460 feet (140 m) from one side of the paleoisland to the other. Based on the elevation of Silurian strata off to the northwest side (fig. 4.3b), it is clear that limestone of one kind or another completely buried the island's original topography, only for the structure to be partially exhumed by erosion in recent times to form a kind of geologic window.

Variations in the dip of strata wrapped around the structure's perimeter may be due not only to natural deviations in the original slope but also to some degree of postdepositional folding on a regional basis. The fact remains that silty limestone and large stromatoporoid fossils occur only on the southeast flank, whereas conglomerate and clean limestone dominate the structure's northwest face. This dichotomy informs us that structural folding alone provides an inadequate explanation for variations in strata on the unconformity surface. Hence, the structure's paleotopography implies a relative rise in Silurian sea level of no less than 100 feet (~30 m). After exploring the island's structure on foot, the question became whether or not comparable changes in Silurian sea level might be recorded elsewhere in strata of the same age. To answer that query, it was necessary to determine the actual age of Silurian strata at Bater Island as accurately as possible. Age can be determined through analysis based on macrofossils preserved in the limestone beds around Bater Island but even more precisely through the recovery of microfossils representing organisms that underwent rapid evolution.

CORALS, STROMATOPOROIDS, AND CONODONTS

Three collection localities on Bater Island have been studied extensively for their macrofossils.[6] From site 1 on the south side of the diorite core, large stromatoporoid colonies (fig. 4.4b) are identified as *Plexodictyon* sp. Other macrofossils are scarce but include two kinds of tabulate corals (*Okopites subtiles* and *Thamnopora* cf. *neimongolensis*). One of these is more compact in form with low-profile colonies (*O. subtiles*), whereas the other (*T.* cf. *neimongolensis)* takes on the growth form of more delicate, finger-like columns. The greatest coral diversity occurs at site 2 (fig. 4.3a), still on the south side of the island but more eastward. In total, seven coral species assigned to six genera are described from this locality (*Mesoculipora* cf. *divida*, *Thamnopora* cf. *neimonglolensis*, *Striatopora* cf. *microsepala*, *Striatopora* sp, *Cladopora obesa*, *Taxopora* sp., and *Planocoenites* sp.). One species (*C. obesa*) is notable for its finger-shaped, branching colonies. In addition, the same locality yields three species of stromatoporoids (*Clathrodictyon gotlandense*, *C. microstriatellum*, and *Actinostromella slitensis*) and a fourth identified only to genus level (*Hexastylostroma* sp.). Again, *Cladopora* sticks out as a columnar or branching type of tabulate coral. The largest such branching coral from this site was found to have a colony diameter of 60 inches (1.5 m) and a height of 15.5 inches (40 cm). No corals were recovered from site 3 on the north side of the island (fig. 4.3a), but two stromatoporoids were identified (*Clathrodicyyon gotlandense* and *Actinostromella slitensis*). Overall, the greatest number of coral and stromatoporoid species is limited to the southeast margin of the island, numbering ten taxa in total.

Regarding geologic age, some of the fossil corals (including *Cladopora*, *Thamnopora*, and *Planocoenites*) are commonly found in Devonian strata but are also known from the Upper Silurian of north China. Generally, much of the fauna is regarded as belonging to the Ludlow series within the upper part of the Silurian system. The corresponding Ludlow epoch lasted about four million years, ending a couple of million years before the start of the Devonian period (see fig. 1.8). The place name for these particular geological partitions derives from outcrops below England's Ludlow castle in the Shropshire town by that name.[7]

Fossil graptolites definitive for the Ludlow series and its two subdivisions (Gorstian and Ludfordian) are unknown in China's Inner Mongolia. As in the case of the limestone-rich area around Churchill in Canada's northern Manitoba (see chapter 2), conodont microfossils are reported from the Joint Banner Dahrhan-Mumingan district of Inner Mongolia. To ascertain a more precise age for the Xibiehe Formation, conodont samples were collected from the base of the formation and upward through layers near Bater Island.[8] The formation's upper layers yield a distinctive conodont species (*Ancoradella ploeckensis*) correlated with the middle of the Ludlow series (Ludfordian stage). On that basis, it is possible to infer global changes in sea level that occurred about 421 million years ago.

WINDWARD AND LEEWARD SHORES

As the sea level rose around the margins of Bater Island on the North China Block, shore erosion and the establishment of marine life in preferred zones along the rocky coast responded to differences in wave shock. In particular, conglomerate consisting of diorite cobbles and small boulders are limited to the paleoisland's northwest flank. The limestone that follows after is clean and lacks any trace of silt. Among the three areas where macrofossils were collected and subsequently studied, the northwest locality at site 3 (fig. 4.3a) was the least diverse.[9] The combination of physical factors indicates that the northwest shore was subject to surf capable of eroding sizable clasts from parent diorite rock on an exposed rocky shore with sufficient energy to remove fine-grained sediments. Under this windward setting, colonization of marine life in shallow subtidal conditions was sparse and limited to two species of stromatoporoids (*C. gotlandense* and *A. slitensis*).

In contrast, the southeast flank of Bater Island records a thin, initial layer of well-cemented siltstone followed by a thick succession of silty limestone.

The preserved silt layer is dense with constituent grains of plagioclase, chlorite, and augite derived directly from the underlying diorite. Wave activity on this side of the island was insufficient to disperse these finely eroded materials. The more westerly site 1 on the south shore features the largest macrofossils, represented by dome-shaped stromatoporoids up to 24 inches (60 cm) in diameter preserved in growth condition (fig. 4.4b). The superior size of these armored sponges compared to smaller colonies of other species on the opposite side of the island implies that growing conditions were stable and sustained over longer time spans in a more sheltered setting. Stalked sea lilies, or crinoids, lived fastened to the rocky substrate among the stromatoporoids. Although the crowns of these echinoderms are not preserved, their abundant crinoid columns remain scattered on or very close to the diorite surface. Some crinoid stem segments are up to 2 inches (4 cm) in length (fig. 4.5). The scenario is consistent with calm conditions and weak water currents unable to sweep away the debris of disarticulated crinoids.

The more easterly site 2, on the southern flank of Bater Island, accounts for the most diverse association of corals and stromatoporoids found on the paleoisland with a combined count of ten taxa.[10] Among the corals, those with delicate finger-like columns (*C. obesa* and *T.* cf. *neimongolensis*) were prone to breakage in turbulent water, and they stood a better chance for sustained growth under more sheltered conditions. Although some coral fronds were subject to fragmentation, the broken fronds occur in clusters showing they were not scattered by currents. The high biodiversity from site 2 is in agreement with the interpretation of a tranquil setting free from much wave agitation. Overall, the dichotomy of environmental indicators on opposite sides of Bater Island suggests a contrast in windward and leeward conditions, with strong surf on the windward side and calm conditions on a protected leeward shore.

PREVAILING WINDS

On any given coastline, sea swells that generate surf and crashing waves are typically wind-driven in origin. How, then, might the perceived variations in wave energy around Bater Island be related to possible patterns of atmospheric circulation in a late Silurian world? In the present-day Pacific and Atlantic basins, subtropical high-pressure zones are centered around the latitudes of 30°N and S of the equator (fig. 4.6), where the southeast and northeast Trade Winds radiate outward to push sea swells toward the Intertropical Convergence Zone (ITCZ). The zone of confluence shifts somewhat north or south on a seasonal

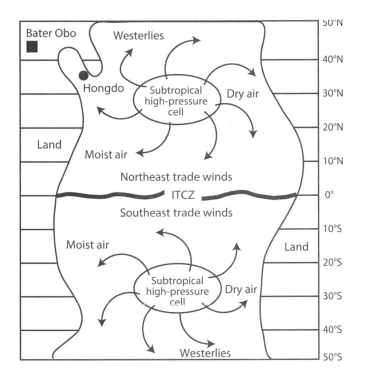

FIGURE 4.6 Present-day scheme for atmospheric high-pressure cells in the subtropics that stimulate the trade winds and westerlies over the Pacific Basin.

basis, but a strong, equatorial current flows from east to west under atmospheric conditions of low pressure resulting from rising air that is also humid. The atmospheric engine that ameliorates land-surface and sea-surface temperatures across latitudes from pole to pole creates the Hadley cells. After moist equatorial air rises high enough in the atmosphere to undergo cooling and releases its water as rain in the tropics, dry air continues to circulate both northward and southward only to sink back to the surface at the so-called doldrums (or horse latitudes), where winds and ocean currents are muted. Prevailing winds at latitudes around 45°N and S of the equator are called the westerlies. They are related to the same high-pressure cells that help create the trade winds that converge near the equator. Continental climates on the margins of the oceans help to strengthen the westerlies as high-pressure cells migrate overland from polar fronts that become more active during the winter season.

The seasonal winter winds that feed the westerlies off the China mainland and out over the Yellow Sea have a profound effect on the erosion and subsequent island geomorphology of Hongdo, the Red Island (see chapter 1). Quartzite cobbles are concentrated as a shore shingle solely on the west side of that island at Mongdol Beach (see fig. 1.2). Today, the same vigorous westerly winds cross the open plains of Outer Mongolia, as well as the mountain at Bater Obo in China's Inner Mongolia (fig. 4.1b). If northern China was in approximately the same location as found today, then the Silurian Bater Island would have been subject to the prevailing Westerly winds that now characterize that part of Asia. However, global map reconstructions with the massive continent of Silurian Gondwana, mostly in the Southern Hemisphere, show that the North China Block was isolated from the South China Block at a latitude still within the Northern Hemisphere but closer to the Silurian equator than today.[11] Much progress has been made since the time of Amadeus William Grabau and his novel paleogeographic maps. The jigsaw puzzle machinations of mapmaking are fraught with compromise constrained by placing the largest paleocontinents with superior geomagmetic indicators first, before adding smaller pieces of microcontinents perhaps lacking any useful geomagmetic information. Such is the case for the North China Block, which likely was located within 10° or 15° of the paleoequator during late Silurian time. If far enough to the south during the Ludfordian age (see later treatment), Silurian Bater Island was subject not to seasonally active westerlies but to the prevailing northeast trade winds at a different latitude.

HISTORICAL VIGNETTE: GENGHIS KHAN
AND HIS MOUNTAIN SHRINE

On the final day of fieldwork in the Dahrhan-Mumingan district, an effort was made to reach the top of Bater Obo. The way was not obvious for motor vehicles, and various false paths were followed and abandoned on the climb to the mountaintop. The summit of Bater Obo registered an elevation of 4,265 feet (1,300 m) based on a readout from one of the vehicles. There, a ceremonial obo was found with three tiers of limestone blocks assembled in a circular pattern, one atop the other like a wedding cake. The structure was adorned by woody brush woven into a protective thicket around the perimeter at each level. A stout pole planted on the top tier was rigged with ropes connected to four corner posts set apart at some distance. One had come loose from its anchorage and lay prostrate on the ground. Prayer flags fastened to the rope lines fluttered in the breeze. An altar adorned with Mongolian script stood at the front of the obo, on which a metal cabinet sat securely cemented in place. The site appeared to be a shrine (fig. 4.7).

FIGURE 4.7 Bater Obo at the mountain-top shrine to Genghis Khan located near the Bater paleoisland.

Photo by author.

Given the traditional shamanistic beliefs of Mongolian peoples, we assumed that the shine bore religious significance in homage to a local mountain spirit.

As we settled down around a picnic blanket to enjoy lunch under a cloudless blue sky, a caretaker immerged from nowhere and began to fuss with the fallen corner pole. The man was no match for the heavy post, which prompted one of our party to rush to his aid. With the spell of apprehension broken, the attendant offered to open the locked cabinet atop the altar. We eagerly crowded around to see within as the doors of the altarpiece swung open to reveal offerings of small cakes, bottles of liqueur, and a nest of Chinese paper currency. It came as a surprise to find innermost a portrait of Genghis Khan. The shrine celebrated no mere mountain spirit but the real-life, thirteenth-century "universal ruler" who forged the world's greatest empire radiating from the heartland of Mongolia to encompass territories from Korea in the east to Persia in the west, to the Volga River area of Russia in the north, and eventually all of China in the south.

Without a father to protect him, the boy Temuchin (1162–1227) overcame many difficulties in adulthood as the unifier of warring Mongol clans and organizer of an unstoppable military machine that reached the gate of eastern Europe.[12] The nomadic Mongols were predisposed to forge a formidable fighting force by dint of horsemen armed with double-curved compound bows and arrows designed for short- and long-range fire, both facing forward and backward with footing supported by iron stirrups. The great Khan's success drew from a combination of administrative planning that awarded meritocracy based on military strategy with a previously unrealized level of communications among field commanders, sound judgment of character, and, not least of all, his unshakable belief in the possibilities of a vast empire.

Departure from Bater Obo on that bright summer day in 1999 was equally challenging in the absence of a clear track down the slopes to the steppes below. The mountain's Silurian capstone gave pause to consider that perhaps igneous basement rock lay hidden below as the foundation of yet another paleoisland. Indeed, the overall topography of the borderlands with Outer Mongolia invoked a ghostly former seascape in a cloaked Silurian archipelago. As an afterthought, the launch of future investigations using seismic studies on these suggestive "island" mounds is a distinct possibility.

OTHER SILURIAN LANDS AND ISLANDS

Swedish Gotland resides in the Baltic Sea of northern Europe, appealing as a gentle island landscape controlled by underlying Silurian stratigraphy. The Wenlock and Ludlow series are well exposed as limestone and sandstone strata that dip regionally from northwest to southeast across a unified structural block with much of the Hemse Group and the overlying Eke, Bergsvik, and Sundra Formations constrained to the Ludfordian stage on the basis of conodonts. As such, discrete paleoislands within the Upper Silurian of Gotland are not readily apparent from geologic mapping, although large coral mounds with considerable topographic relief are a common feature. On the island's east shore at Kuppen, however, Ludfordian strata exhibit exposures of Upper Hemse beds that represent a rocky coastline exposed during a drawdown in sea level. A distinct row of limestone sea stacks aligned over 650 feet (~200 m) was eroded during this interval and subsequently buried by beach gravel during the following rise in sea level.[13] The gravel was derived from parent limestone exposed in the sea cliff. It consists of pebble-sized clasts eroded as fossil fragments of stromatoporoid sponges mixed with crinoid debris. The maximum relief of these Silurian sea cliffs is no more than 6.5 feet (2 m). Rare examples of encrusting tabulate (auloporid) corals are preserved in their original growth position on the erosion surface.

The Kuppen localities are bracketed in time by older and younger strata from which fossil chitons have been recovered.[14] Characterized by disarticulated plates (or sclerites), the chiton fossils were not recovered from a rocky-shore setting but lived under rough-water conditions on a shallow, carbonate ramp. The eight-plate configuration typical for chitons is envisioned in the ecological reconstruction of these Silurian mollusks. The largest such polyplacophoran (*Chelodes actinis*) was as much as 4.75 inches (12 cm) long based on an estimated width-to-length body ratio. As many as six chiton species in as many genera are recognized from the Upper Silurian of Gotland. The ancestors of Silurian chitons and oldest known representatives occur in Upper Cambrian strata, widespread at localities ranging from Utah, Wisconsin, and New York.[15] The parent chiton species (*Matthevia variabils*) possessed spiny plates to deter predation and is interpreted as a grazer living on stromatolites under intertidal conditions. Stromatolites are formed by the build-up of filamentous bacteria, not to be confused with the sponge-like armored stromatoporoids common as fossils from the Upper Silurian of China's Inner Mongolia and Sweden's Gotland. Chitons may have occupied the rocky shores of Wisconsin's Baraboo islands (see chapter 2), but separate sclerites have yet to be found.

LATE SILURIAN GLOBAL GEOGRAPHY

As during the Ordovician period (see Fig. 3.9), the distribution of Silurian lands was dominated by the supercontinent of Gondwana, which consisted of a single landmass formed by present-day South America, Africa, Arabia, India, Antarctica, and Australia, as well as smaller fragments such as Arabia and parts of southern Europe located almost entirely in the Southern Hemisphere (fig. 4.8). Today's China was not a coherent land but separated as three microplates represented by the South China Block (Yangtze platform), North China Block (including Inner Mongolia and Korea), and western China (Tarim). These entities were loosely clustered to form a blind alley, with south China straddling the paleoequator, and north China and Tarim, located north of the paleoequator. Under this configuration, a vigorous long-shore current was driven by prevailing westerly winds at high latitudes along the shores of South America, North Africa, and southern Europe, reaching the Yangtze platform at a low latitude where a change over to the southeast trade winds energized an equatorial current brushing the shores of the North China Block and Tarim on the way to Baltica (western Russia and Scandinavia). Compared to its earlier place in Late Ordovician time, Baltica

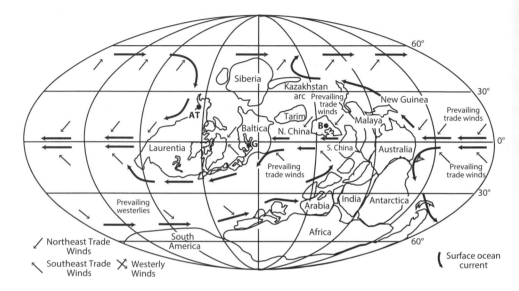

FIGURE 4.8 Global reconstruction for the Late Silurian (Ludlow) world with interpretations for trade winds convergent on the paleoequator, high-latitude westerly winds, and surface ocean circulation. Abbreviations: B = Bater Island (North China Block); G = Gotland Island (Sweden). Modified and expanded from the reconstruction by Johnson et al. (2004, their fig. 7).

tracked as much as 25° of latitude northward toward the close of the Silurian period such that Sweden's Gotland drew closer to the paleoequator.

Notably, Upper Silurian strata from Bater Island and Gotland yield the same stromatoporoid species (i.e., *C. gotlandense* and *A. slitensis*). It may be argued that the reproductive propagules of these organisms were seeded westward from the shores of one continent to another by the equatorial current. The case may be made that Silurian Bater Island was much closer to the equator than found today poleward of 42° north latitude, where winter westerly winds are prevalent in that part of Asia. Circulation driven by westerly winds around the same latitudes in that part of the Silurian world pushed a completely different ocean gyre through a northeasterly and westward current in the opposite direction from Scandinavia. A significant tectonic development during the latter part of the Silurian period was the ongoing closure of the Proto-Atlantic (Iapetus) ocean between Baltica and ancestral North America (Laurentia). The collision between Greenland and Norway was about to occur during the ensuing Devonian period.

A separate issue concerns suspect terranes that undoubtedly occurred as individual islands independent from the principal Silurian continents. The Alexander terrain (fig. 4.8) is one of more than a dozen island terrains that collided with

Laurentia that can be dated to the latest Silurian and earliest Devonian.[16] It is unknown how far away these wayward archipelagos may have been located off the westward coast of Laurentia before they joined the larger land mass.

SUMMARY

Bater Island is petite compared to other paleoislands from the older Baraboo and Jens Munk Archipelagos, but the human visitor to its Silurian heights can all but hear the muffled cacophony of waves breaking on a cobble-shingled coast in the windward direction and sense the tranquility of a sheltered shore on the opposite coast. The spot lacks any sign of human habitation, and the explorer may be forgiven for sensing the echoes from the deep past just as plainly as if standing on a remote wind-blown and sea-battered island in the modern world. The rhythms of the sea's antiquity with pounding surf and the flow of ocean currents are brought alive. In a place like this, the spirit of Rachel Carson is easily conjured. Like her, we can feel the wind in our hair and hear the sound of waves on a long-vanished cobble beach. Places as distant from one another as the geological window into Bater Island in the shadow of Bater Obo on the grasslands of Inner Mongolia and Sweden's bucolic Gotland are but pinpricks on a global map. Yet, they are linked by the strands of time that speak to anyone who would consider changing sea levels on rocky coastlines as well as the colonization and transferal of life from one seashore to another over vast distances on the global stage. Biological actors come and go with time, fated by extinction, but the play remains the same scene-by-scene in its physical plotline.

CHAPTER 5

HOW BIGGER ISLANDS ARE BROKEN INTO SMALLER PIECES

A Journey in Late Devonian Time to Western Australia's Mowanbini Archipelago

> *Here and there a lonely tower of rock rises offshore, one of the forma-tions known as stacks or needles. Each began as a narrow headland jutting out from the main body of coastal rock. Then a weak spot in its connection with the mainland was battered through.*
>
> —Rachel Carson, *Our Ever-Changing Shore* (1958)

The island continent of Australia is a place of superlative attrac-tions, possessed of richly varied ecosystems that draw thousands of ecotourists every year. Stretching some 1,250 miles (~2,000 km) in length, the Great Barrier Reef off the Pacific coast of Queensland thrills visi-tors with its mix of hard and soft corals dressed in vibrant hues of red, green, and blue. Onshore, the Queensland rainforest features giant ferns, conifers, and flowering plants with historical roots that trace back to the time of dino-saurs. Western Australia on the Indian Ocean is renowned for the Shark Bay World Heritage Area. The living stromatolites there represent the planet's oldest macrostructures, which first appeared as microbial growths billions of years ago. Ashore, the coastal desert is carpeted by wildflowers during the austral spring that appeals to many nature enthusiasts. Uluru (Ayer's Rock) is the iconic monolith near the continent's center, placing aboriginal people in the middle of an ecosystem that underscores time-honed survival skills under harsh conditions. Enthralling as these ecosystems are, Australia is home to an array of fossil ecosystems preserved on a physical scale just as impressive as their living counterparts.

The Devonian period (see fig. 1.9) represents one of the most profound biological transitions on planet Earth with ramifications both on land and at sea. The terrestrial greening of the planet occurred by the Late Devonian when tree-size plants spread inland to build forests dominated by giant lycopods (club mosses), horsetails, and ferns. These were the earliest vascular plants equipped with sturdy internal plumbing capable of circulating water and food sugars to all parts. The Australian *Baragwanathia* flora of lycophytes made an appearance in the Southern Hemisphere before spreading to the rest of the world.[1] Also known as the age of fishes, Devonian faunas from Australia are diverse in armored fish (placoderms), sharks, and lobe-fin fish (Sarcopterygii) equipped with bony joints that prepared the way for colonization of the land by amphibians and reptiles.[2] Both kinds of four-legged creatures (tetrapods) were ascendant during the Devonian.

Moreover, the Devonian also was when barrier reefs achieved the status of coastal ecosystems. Western Australia's Canning Basin in the Kimberley district (fig. 5.1a) is home to the largest such fossil reef in the world, with a continuous exposure of more than 200 miles (~350 km) adjacent to the continental Kimberley Block. The actual length of the reef tract was much greater than now exposed and may have rivaled Queensland's Great Barrier Reef in scale. What remains is called the Devonian Great Barrier Reef, the geology and paleontology of which have been studied and mapped in detail.[3] Separated by 62 miles (~100 km), two of Australia's national parks capitalize on cross-sections through the same fossil reef. In the north, the Windjana Gorge National Park features a 2-mile (3.5-km) trail along the Lennard River, cutting through towering cliffs to expose back-reef, reef-core, and fore-reef deposits. Geikie Gorge National Park is farther to the southeast and facilitates popular boat tours through the reef on the Fitzroy River.

This chapter treats the reef segment skirting outward from the granite and metamorphic terrain of the Kimberley Block around the Oscar Range (figs. 5.1b; 5.2a). Paleoproterozoic rocks exposed in the range predate the Devonian period by roughly two billion years. They are composed of interbedded quartzite and softer phyllite layers deformed by complex folds. Much as found today, the rocks of the Oscar Range were exhumed in Devonian time to stand as monadnocks in the style of New Hampshire's Mount Monadnock. During Late Devonian time, however, they were drowned in the image of South Korea's Hongdo (Red Island). Resistant to erosion, the pervasive quartzite layers within the range continue to uphold those same monadnocks today. The Oscar Range extends over 68 square miles (175 km²), approximately three-quarters of the size preserved in the Cambrian Baraboo Archipelago of Wisconsin (chapter 2). The prominence of the Oscar Range amounts to nearly 300 feet (~90 m), slightly more than the

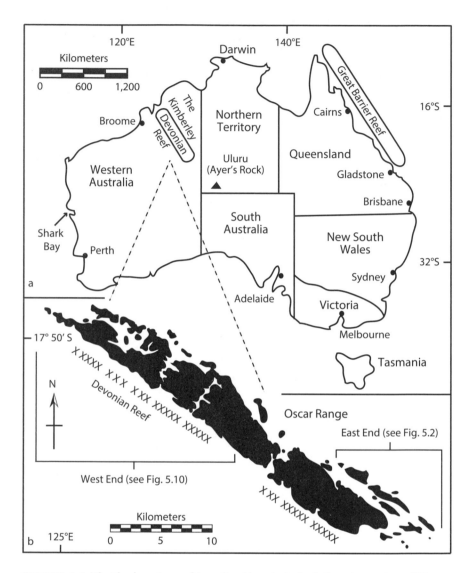

FIGURE 5.1 The island continent of Australia with geologic details from the province of Western Australia: a) major Australian cities and notable natural features such as the Great Barrier Reef, Uluru (Ayer's Rock), and fossil Devonian reef in the Kimberly district of Western Australia; b) detailed outline of the Oscar Range showing its relationship to a part of the Great Devonian Reef.

exposed prominence of the largest Baraboo monadnock. The eastern end of the range is readily accessible by road and features a half-dozen smaller islands on the inner flank, a large island, and a well-exposed reef on its outer flank (fig. 5.2b). Adopting the original place name from the aboriginal Bunaba people, whose

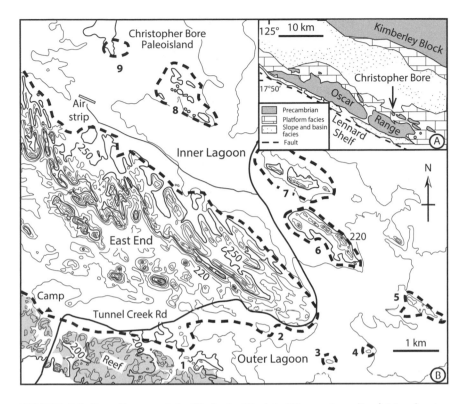

FIGURE 5.2 Oscar Range and the Kimberley Block in Western Australia. a) Map showing general relationships between the older (Precambrian) rocks of the Oscar Range and surrounding Devonian limestone facies; b) detailed topographic map covering the east end of the Oscar Range together with Devonian reef rocks and related outer- and inner lagoon limestones.

ancestors arrived in Australia roughly sixty thousand years ago, the formal designation Mowanbini Archipelago was conferred on the two largest paleoislands, more than twenty satellite islands, and countless islets and sea stacks arrayed around the margins of the Oscar Range.[4]

FEET ON THE GROUND

Leaving Bater Island behind on the North China Block, our geological clock is recalibrated for an advance in time by forty-five million years to set foot on a planet Earth that existed 375 million years ago. During that time, the loose association of stony stromatoporoids that grew in place on the shores of Bater

Island evolved into an astonishing proliferation of species knitted together to construct a true reef framework, as well as others more suited for life in sheltered lagoons. We have reached a nodal point near the end of the Frasnian age in the Late Devonian. With an unsteady hand setting the geologic clock only a million years later, we would witness a multitude of specialized stromatoporoids and corals succumb to the blue planet's second great mass extinction. A cascade of worsening events is linked to black shales that signal widespread deoxygenation of shallow marine waters.[5] Life in tropical seas was more gravely affected than elsewhere at higher latitudes. Stagnant, anoxic water appears to have poisoned tropical organisms with hydrogen sulfide, perhaps due to algal/bacterial blooms. Our landing place, however, puts us squarely at the inner margin of a massive reef structure before the Devonian mass extinction.

An excursion out and back from camp before breakfast leads to the nearest landfall a mile (1.6 km) away. Illuminated under the morning sun (fig. 5.3), the view seaward from a knoll could not be more dramatic in composition. Dipping steeply

FIGURE 5.3 View looking south from an outcrop of Paleoproterozoic bedded quartzite across flats underlain by Paleoproterozoic phyllite towards dolomitized Devonian reef limestone just below the horizon. Person for scale.

Photo by author.

70° to the southwest, a rim of bedded quartzite rises 213 feet (65m) above ground, distinguishing it from the opposite ridge of Devonian reef rock originally formed by limestone. The intervening flats are covered in grass and scattered eucalyptus trees, but loose surface materials hint at phyllite bedrock. Like quartzite, phyllite is another metamorphic rock, but its derivation comes from shale by way of mud instead of sandstone via sand. The linkage with bedded quartzite remains to be seen, but these phyllite flats clearly belong to the metamorphic Oscar Range. The nearby reef margin sits on the outer edge of the flats. Limestone may or may not have been deposited in a lagoon atop the phyllite, but where the flats narrow, as here, the reef margin has the feel of a fringing reef.

Back at camp, breakfast is consumed. In preparation for the day's outing, water bottles are filled and lunches are packed. A fresh line of march follows eastward along cliffs of reef rock (fig. 5.2b). The route passes through low grass past a line of bottle-shaped boab trees (*Adansonia gregorii*), still leafless in the dry season (fig. 5.4a). Passing a vast stone quarry, open cuts in the reef rock provide another perspective. But after nearly a mile and a half (~2 km), direct evidence of the anticipated unconformity between Paleoproterozoic phyllite and overlying limestone remains elusive. The first clue is met in a clearing beyond the quarry, where patches of phyllite protrude above the ground in thinly bedded layers to exhibit a unified dip to the southwest (fig. 5.4b). The trend is consistent with the bedded quartzite on the other side of the valley to the north. Farther along, a cavity below the reef rock reveals the contact with underlying phyllite. It would be easy to miss without paying close attention. Dropping into a streambed below the level of the limestone, the embankments on both sides cut into near-vertical layers of phyllite. The south bank is covered by vegetation that hides the contact with the overlying reef rock.

Former Delta System

Two miles (~3 km) out from the start of our easterly trek, the trees that arch over the stream bed open to grassland at a bend in the embankment (locality 1, fig. 5.2b). Shockingly red to ochre in color under the bright sunshine, an outcrop of conglomerate with abundant quartz pebbles and platy fragments of phyllite offers a dramatic change of scene (fig. 5.5a). The reddish color is due to iron-stained sand forming the matrix within which pebbles from white vein quartz are encased together with broken bits of phyllite (fig. 5.5b). The quartzite and vein-quartz pebbles are angular to subrounded.[6] The deposit also includes a few

FIGURE 5.4 Great Devonian Reef landscapes: a) inner reef margin represented by massive cliffs of dolomitized limestone (boab trees for scale, 16 ft or 5 m tall); b) surface erosion close to the inner reef margin with dip angle of Paleoproterozoic phyllite exposed above the surrounding ground.

Photos by author.

FIGURE 5.5 Deltaic outwash conglomerate against the inner margin of Devonian reef strata: a) view south on outcrop (hammer for scale with 15-in or 38-cm long handle); b) Close-up view showing larger clasts derived from vein quartz (white) and broken bits of platy phyllite (pocket knife for scale, 3.5 in or 9 cm).

Photos by author.

fossil stromatoporoids and marine snails (gastropods). As much as 6.5 feet (2 m) of the distinctive conglomerate occupies the base of a commanding package of strata rising to the south.

The significance of the conglomerate cannot be overstated. The source of the sand, quartzite pebbles, and vein-quartz pebbles derives from the quartzite hills across the valley to the north, whereas the phyllite fragments were ripped loose from the intervening flats. Overall, the conglomerate represents an outwash deposit transported by stream flow out of the quartzite hills and across the phyllite flats.[7] Larger bits of vein quartz captured in the flow had insufficient time to become smooth and well-rounded by abrasion, which suggests short-lived flood events due to torrential flow from rain storms. Marine fossils mixed into the conglomerate point to a shoal-water delta pressed against the inner margin of a fringing reef from which the marine organisms were scoured.

A rock column of more than 98 feet (30 m) rises above the basal conglomerate to reveal layers of muddy to sandy limestone capped by 20 feet (6 m) of massive limestone that includes large stromatoporoids and coral colonies (*Disphyllum* sp.) typical of the reef margin.[8] Some bedding surfaces feature oncoids as much as two and three-quarter inches (7 cm) in diameter and closely packed together (fig. 5.6). These are spherical growths formed by concentric addition of thin calcium-carbonate layers secreted by microbial bacteria and blue-green algae. Locked in solid stone, the oncoid's robust presence all but audibly announces the back-and-forth agitation of waves washing through and around the reef margin.

Former Sea Stacks and Skerries

The day's progress is measured by the sun's approach to its zenith. Leaving behind the Devonian reef complex and delta, the distance across the flats north to the closest quartzite island is nearly a mile (~1.5 km) off. The dominant vegetative groundcover is the tough, spiky tussock grass called spinifex (*Triodia* sp.), which grows in clumps and jabs at ankles. The kapok tree (*Cochlospermum gillivraei*), with its bright yellow flowers, adds a splash of color. Here and there, amongst the vegetation, flagstones of carbonate rock are exposed. They show that shallow marine waters flooded this space between the inner reef margin and islands of the Oscar Range. We may be wading through the spinifex in a desert landscape, but simultaneously, we wade through a shallow Devonian lagoon. Good exposures of phyllite signal our arrival at the paleoshore. The contact between

FIGURE 5.6 Dolomitized Devonian limestone closely associated with the inner reef margin showing census grid used to tabulate fossil density (inset shows typical oncoid in cross section).

Photo by author.

spotty limestone and more persistent phyllite can be traced eastward now until quartzite layers rise above ground (locality 2, fig. 5.2). Climbing onto a massive quartzite ridge, which dips steeply to the southwest, a lunch spot is secured from which we may look back on our path from the inner reef margin.

At closer range, an area of about 35 acres (~141,500 m²) features an abrupt bend in an island coastline that cuts through the quartzite ridge to define an embayment open to the southeast (fig. 5.7). A scree slope covered with broken pieces of quartzite falls some 17 feet (15 m) from the end of the ridge to meet flat terrain with flagstones that look much like dark rain puddles in a sea of grass. There is much to be explored, and lunch is quickly consumed. Whereas the ridge crest is upheld by a single massive unit of quartzite, the landscape falling to the north and south is formed by interbedded layers of phyllite and quartzite. Descent from the nose of the ridge to the paleoshore brings us laterally across some 200 feet (~60 m) of phyllite interspersed with slender bands of resistant quartzite that project upright in places like tombstones.

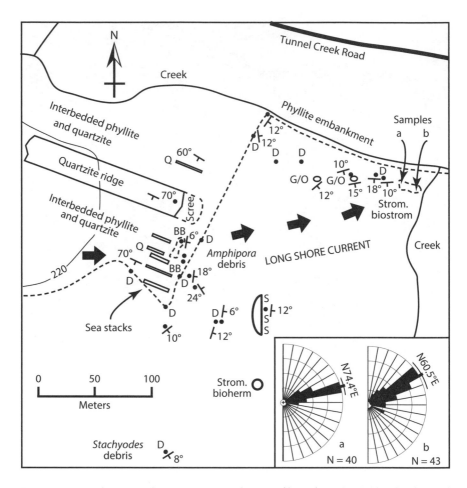

FIGURE 5.7 Paleogeographic reconstruction for site 2 (fig. 5.2) on the south side of Tunnel Creek Road (dashed line indicates shoreward limit of dolomitized limestone; Q = quartzite; BB = basal breccia; D = bedded dolomite; G/O = gastropods and oncoids; S = spongiomorphs). Inset shows rose diagrams that plot the orientation of stromatoporoids from two sample levels (a and b on map). Modified from Johnson and Webb (2007, their fig. 4).

Fractured quartzite includes abundant vein quartz stained red from leached iron compounds much in the same fashion as encountered at the tide-water delta on the reef margin, now in the far distance across the flats. Here, the rubble includes larger pieces of quartzite that were eroded directly from the adjacent paleoshore.[9] Realistically, the quartzite tombstones with their breccia halos signify former sea stacks on the coast of a Devonian lagoon (fig. 5.7), where wave and current action gradually wore away the interbedded phyllite with sufficient energy to also degrade the standing quartzite layers. The lesser tombstones are nearly awash in rubble, and their reduced relief brings to mind the more obscure term for skerries, or low rocky islets found along rugged rocky shores in places like Scotland.

The shoreline's inflection is apparent from the observation that all the tombstones, large and small, are truncated to the east. These fall in line with the scree slope off the nose of the main quartzite ridge. The paleoshore passes beyond the quartzite ridge as far as 330 feet (~100 m) to bypass low topography dominated by phyllite. Standing on the flats in a semidesert setting far from the nearest shores of the Indian Ocean, it is a stirring sensation to imagine the persistence of a Devonian coastal current capable of eroding not only the multiple sea stacks but also the more stalwart quartzite ridge on which we took lunch. To confirm the dynamics of such an extraordinary spot captured in Devonian time, we wade into the shallows to find how marine life responded to the wind and waves.

Scattered flagstones of dolostone laid bare to the sky afford a direct window onto the Devonian seabed. Closer to the outermost sea stack, the flagstones show no fossils but record an offshore dip of 10°. Farther south, strata show a lesser dip of 8° in the same direction (fig. 5.7) and reveal a bedding surface with the branch-like remains of *Stachyodes* (fig. 5.8a). Like a thicket of coarse twigs standing upright in the water by as much as 7 inches (12.5 cm), the invertebrate was a peculiar kind of stromatoporoid that favored an open lagoon environment. At such a distance offshore, water depth was sufficient enough that clusters of *Stachyodes* were not subject to rough water to the same degree as the coastal sea stacks. A larger patch of exposed flagstone more to the northeast exposes an assemblage of pillar-shaped stromatoporids of the kind previously found at the reef margin. These are preserved upright in growth position, and their relationship in a dense heap indicates a small bioherm (or life mound).

Inshore from the bioherm, flagstones with extensive assemblages of calcified spongiomorphs having a spherical growth pattern are well exposed in layers that dip 12° eastward in an offshore direction. Continuing northward into the embayment but closer to the paleoshore, flagstones with fragmented *Amphiopora* are increasingly common. These represent yet another variation on the theme of stromatoporoids, in this case, twig-like in form but much more delicate than *Stachyodes*. Reaching the farthest incursion of the embayment at its northwest corner, flagstones of dolomitized limestone dip 12° eastward, as before. After a sharp bend in the paleoshore, the layers change dip direction to slope more southward following the contours of the bay but, on average, still 12°. Here, bedding surfaces reveal clusters of low-spired marine snails (gastropods),

FIGURE 5.8 Devonian fossil associations from a lagoon setting: a) robust, branch-like morphology of peculiar stromatorporoids called *Stachyodes* (pencil for scale); b) small oncoids encrusted around marine snails (gastropods).

Photos by author.

encrusted by microbial growth as oncoids. Some amount of energy in the water column was necessary to induce the complete encrustation of the snail shells by microbial growth.

Overlapping layers with concentrations of pillar-shaped stomatoporoids occur scarcely 30 feet (~10 m) off the bay's inner embankment formed by phyllite. One species (*Stromatopora* cf. *cooperi*) is dominant, but at least two others also are present (*Clathrocoilona spissa* and *Hermatostroma ambiguum*). The assemblages are biostromal (or flat layered) in arrangement and follow parallel to the paleoshore. Individual fossils are up to 6 inches (15 cm) in diameter but are not preserved in an upright growth position. Rather, the colonies are uniformly tilted in a northeasterly direction as recorded by compass bearings from two levels of strata (fig. 5.7, see inset). Under this scenario, insights from the bedded stromatoporoids are consistent with the eroded sea stacks elsewhere on the paleoshore. A vigorous current swept eastward along this part of the shore, influencing stromatoporoid growth in a preferred direction.[10]

At locality 2 (fig. 5.2b), the inner part of the Devonian shoreline strikes parallel to Tunnel Creek Road against a phyllite shoreline before curving around the east end of the Oscar Range, with its heavy quartzite highlands. The afternoon is advanced, but we press onward across the flats to the southeast to diverge from the main paleoshore toward a low knoll on the horizon. The goal is a knob of Paleoproterozoic quartzite that rises ever so slightly above the surrounding limestone landscape as a former islet (locality 3, fig. 5.2b). Looking back toward the quartzite hills of the Oscar Range, it is clear that the islet is aligned with one of the prominent ridges in Paleoproterozoic basement rocks. Regrettably, the Devonian flagstones in this part of the lagoon no longer join against the older basement rocks. A moat of fine, red dust encircles the islet.

Former Outlying Islets

A new line of march is set in a north-by-northeast direction, and we find ourselves midway between the east-most hills on the nose of the Oscar Range and another inlier of basement rock farther off to the east (locality 4, fig. 5.2b). This medial position shows that the Paleoproterozoic knoll to the east is an islet in alignment with one of the thickest quartzite ridges in the Oscar Range. Devonian flagstones are plentiful and expose bedding planes that exhibit concentrations of fossils, like the twiggy *Amphipora*. A halt is called to closely examine on hands and knees a rock surface crowded with snail (gastropod) shells encrusted by oncoids (fig. 5.8b). The largest is barely 5/8 inch (1.5 cm) in diameter, including the microbial rind. Once again, the rhythm of waves and currents scouring the shallow seabed leaps from the silent rocks. The marine snails lived independently and expired before their shells became the hard nucleus for the colonizing oncoids. Near-constant agitation of the water column was a prerequisite for the growth of each oncoid to form a complete rind.

Other paleoislands (localities 4, 5, and 6, fig. 5.2b) are in sight, located to the east and north. One can imagine a time more remote when the Oscar Range extended unbroken farther east and southeast. Eventually, the sea broke through

thin quartzite headlands that linked future islets with the greater island mainland. As hinted by Rachael Carson, we bear witness to the diminution of stout
quartzite islands into smaller bits and pieces left in place as peripheral islets.
With the sun sinking closer to the horizon, a return to camp is in order. A course
westward across the flats intersects Tunnel Creek Road. Progress is faster over a
3-mile (~5-km) stretch on the road. Tilted quartzite layers form ridge lines that
run parallel to the road to the north. Lengthening shadows spill over the former
Devonian lagoon to the south. As the heat of the day dissipates, wildlife ventures
out into the open. Be prepared to see a kangaroo hopping across the flats with the
great Devonian reef as a backdrop.

MORE ON THE LATE DEVONIAN EXTINCTIONS

Timelines that delimit the rise and decline of reef organisms and other life forms
during the Devonian period are characterized worldwide by formal stages that
divide the Upper Devonian into two parts. The Frasnian stage (see fig. 1.9) has a
Global Stratotype Section and Point (GSSP) in the Noire Mountains of southern
France, although the namesake for the stage and its corresponding time interval
derives from the town of Frasnes in the classically studied Ardennes region of
Belgium.[11] Like many other Paleozoic divisions, the base of the Frasnian stage is
defined by abundant microfossils called conodonts. During the Frasnian age, the
great Devonian barrier reef and associated back-reef facies reached peak diversity
in Western Australia.

Reef-related biotas continued to exist during the following Famennian age,
but the diverse stromatoporoids that populated the reef margins and back-
reef lagoons with pillar-shaped stromatopoids (like *Stromatopora cooperi*) and
branching stromatoporoids (like *Amphipora* and *Stachyodes*) suffered extinction with few surviving species.[12] It was the same for tabulate corals (like
Disphyllum sp.) that fell into decline with few progenitors to continue into
Famennian time. A distinct unconformity between Frasnian and Famennian
limestone is exposed in parts of the Canning Basin, including the west end
of the Oscar Range. Much of the Frasnian strata in Western Australia goes
by the name Pillara limestone, whereas the Nullara limestone is the name
applied to succeeding Famennian strata. In stark contrast to Pillara reef margin
and lagoonal limestone, the later Nullara strata are dominated by microbial
mats, oncoids, and bryozoa, filling the former ecological space of extinct stromatoporoids and corals.[13]

OUTER AND INNER LAGOON DYNAMICS

The great Devonian reef around the Oscar Range changes from a fringing reef in the west to a barrier reef in the east. Based on exposures of Pillara limestone between the reef margin and the Oscar Range with its related rocky outcrops, an outer lagoon was flooded in Frasnian time that covered at least 48 square miles (125 km²). A vigorous longshore current eroded quartzite sea stacks on the south coast of the large east island and induced current-oriented growth in pillar-shaped stromatoporoids close to shore (fig. 5.7). Shoal water with energetic bottom currents swept back and forth across a narrow threshold between the eastern end of the Oscar Range and outlying islets, where the seabed was occupied by microbial oncoids nucleated around shells (fig. 5.8b). A grand inner lagoon accumulated extensive deposits of Pillara limestone (and later Nullara limestone) between the north side of the Oscar Range and the continental mainland embodied by the Kimberley Block (fig. 5.2). The area encompassed by the inner lagoon amounts to no less than 193 square miles (500 km²), based on the detailed mapping by Playford, Hocking, and Cockbain.[14]

The eastern islets define a geographic boundary between the outer and inner lagoons flanking the south and north sides of the Oscar Range (localities 3 and 4, fig. 5.2) but offer little insight into the imprint of biological patterns. In contrast, the Christopher Bore paleoisland is among the most revealing spots to characterize marine dynamics in the inner lagoon (locality 9, fig. 5.2b). Situated 1.5 miles (~2 km) off the north flank of the Oscar Range, the tiny paleoisland was the focus of a separate study.[15] It is reached by a side track off Tunnel Creek Road that ends at a watering yard for cattle. On approach from the south, the islet is barely perceptible as a gentle rise on the horizon, but the flats are dotted with flagstones exposed among the spikey spinifex (fig. 5.9a). The islet's shore is abrupt, with topographic relief expressed principally by Paleoproterozoic quartzite beds that are steeply tilted (fig. 5.9b). At best, the Devonian islet strikes a prominence of 6.5 feet (2 m) above the surrounding countryside. Pillara limestone crops out within 1 to 2 feet (15–30 cm) of the paleoshore with thin layers of limestone interspersed with quartz granules (fig. 5.9c). The difference in apparent water energy at the shoreline could not be more striking than observed on the opposite side of the Oscar Range among the quartzite sea stacks (fig. 5.6).

The entire paleoisland is only 62 acres (0.25 km²), but the west end consists of low-lying phyllite. Moreover, soil cover conceals the precise shoreline on the north side. Pillara limestone dips offshore between 10° and 26°, with bedding exposed parallel to the paleoshore on three sides. Survey work based on a series

FIGURE 5.9 Details of outlying Christopher Bore Paleoisland on the inner (north) flank of the Oscar Range: a) wiew across the flats that expose Pillara Limestone to the islet's south shore on the horizon (figure for scale); b) view eastward along the islet's south shore with bedded quartzite rising abruptly with physical relief (figures for scale); c) thin limestone layers with stringers of quartz granules exposed adjacent to the paleoshore (pocket knife for scale = 3.5 in or 9 cm).

Photos by author.

of transects that run perpendicular to the paleoshore confirm a succession of distinct zones roughly parallel to one another.[16] From onshore to offshore, the zones show changes in strata containing laminated limestone lacking fossils, abundant *Amhipora* and *Stachyodes*, pillar-shaped stromatoporoids preserved in an upright growth position, followed by beds with abundant mollusk shells. Layers with

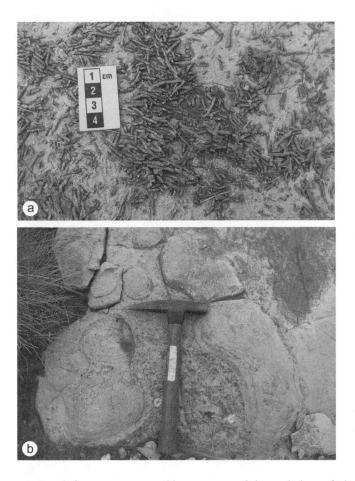

FIGURE 5.10 Details from stromatoporoid biozones around the south shore of Christopher Bore Paleoisland: a) oriented *Amhipora* fragments as debris from an inner zone; b) pillar-shaped stromatoporoids in growth position (hammer for scale with 15-in or 38-cm long handle).

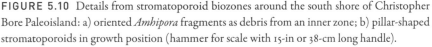

Photos by author.

dense concentrations of *Amhipora* (fig. 5.10a) are mostly constrained to an inner zone landward from the adjacent pillar-shaped stromatoporoids (fig. 5.10b), although *Stachyodes* also occur seaward from that zone. The outer zones with mollusks include separate assemblages of bivalves and gastropods.

Aside from pillar-shaped stromatoporoids preserved in growth position, other biozones feature smaller fossils that were fragmented and moved around on the seabed to a greater or lesser extent. Flagstones with extensive surface exposures of these fossils made it possible to collect hundreds of compass measurements testing the degree to which certain organisms became oriented in consistent patterns after death by waves and current activity.[17] Compass plots on the postmortem orientation of *Stachyodes* fragments show, for example, how those remains came to rest with orientations effectively parallel to the paleoshore (fig. 5.11). Likewise, fossil bivalves from an unknown clam species with an elongated body shape exhibit disarticulated shells oriented roughly parallel to the shore. In contrast, the elongated spires of marine snails attributed to the *Murchisonia* sp. proved to be more random in alignment at four outer stations spread around the paleoisland on three sides.

Wave and current activity within the Devonian inner lagoon around the Christopher Bore paleoisland were not as energetic as in the outer lagoon, where vigorous surf promoted the growth of large oncoids at the inner reef margin (fig. 5.6), and sea stacks were eroded on the opposite rocky shore (fig. 5.7). However, the shores of the Christopher Bore islet were sufficiently scoured by tides or currents to wear away silica grains from the parent quartzite body (fig. 5.9c). Twig-like fragments from broken *Stachyodes* became preferentially oriented with long axes parallel to the shore. The source of waves may have been

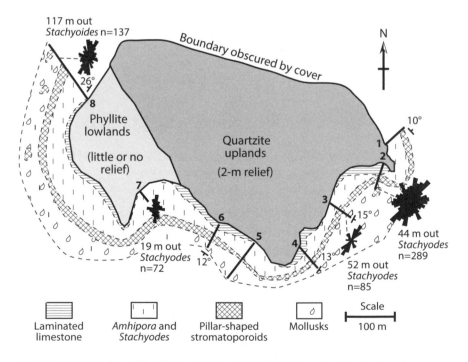

FIGURE 5.11 Outline of the Christopher Bore Paleoisland, located within the inner lagoon on the north side of the Oscar Range, showing arrangement of biozones parallel to the paleoshore with details on compass orientations of *Stachyodes* fragments.

wind-driven, given the fetch between the islet and the north side of the Oscar Range (see fig. 5.2b), but wave refraction around the east end of the range across the threshold between the outer and inner lagoons also affected the paleoisland. In particular, the east side of the islet accumulated more debris from *Stachyodes* than elsewhere along the paleoshore (fig. 5.11).

ONSHORE AND OFFSHORE WINDS

Atmospheric circulation over the great Devonian reef, the lagoons around the Mowanbini Archipelago (i.e., Oscar Range), and the adjacent continental lands of the Kimberley Block was likely to have been regulated by a daily regime of onshore and offshore winds. With the development of an extensive reef system, the Canning Basin clearly was positioned in a tropical setting, where resident organisms found conditions to precipitate calcium-carbonate in enormous amounts. Depending on the latitude of Western Australia in the context of the greater Gondwanan paleocontinent during Frasnian time, vast exposures of igneous and metamorphic rocks on the Kimberly Block and associated hinter-lands were certain to experience substantial changes in ground temperature on a diurnal basis. The Kimberley Block may have been large enough to create its own weather system. During daylight hours, the land mass could be expected to heat up. Warm air would rise upward, attracting an onshore, landward breeze off the adjacent ocean. With nightfall, the land surface would lose heat through rapid radiation, and the warmer ocean water nearby would be expected to attract an offshore, seaward breeze. In concert with daily tides, such a cycle of shifting winds would operate more or less perpendicular to the geographic orientation of the reef and continental coastline. Essentially, the size and thermal capacity of the Kimberley Block at the most propitious latitude may have been sufficient to power wind-driven waves and island currents in regular operation through both the inner and outer lagoons of the Mowanbini Archipelago.

ISLAND DRAINAGE PATTERNS

Coastal rain storms embody another factor that impacted the great Devonian reef, the Mowanbini Archipelago, and nearby Kimberley Block. Today, this is so for Queensland and its associated Great Barrier Reef off the northeast coast of

Australia (fig. 5.1a). Cyclone Debbie became a Category 4 storm that struck the north-central coast of Queensland on March 27, 2017, packing sustained winds clocked at 110 mph (175 k/h). Torrential rain affected large parts of Queensland and tracked farther south into New South Wales. The region's river systems overflowed, causing extensive flooding. Genuine cyclones may have been uncommon over the Canning Basin, positioned on a west-facing shore, but evidence for floods on the Kimberley Block during Late Devonian time is evident from major conglomerate deposits mapped as alluvial fans and fan deltas, several of which occur around the region of the Geikie Gorge. For the most part, these are thought to have been triggered by landslides on Devonian fault scarps, but geomorphic patterns also imply transport abetted by moving water.[18]

Present-day drainage systems are traced with ease throughout the Oscar Range, and a residual system is linked to the Devonian tide-water delta that abuts the reef margin off Tunnel Creek Road (locality 1, fig. 5.2b).[19] Episodic rainfall over the larger islands of the Mowanbini Archipelago was fundamental to the formation of small deltas with sediments derived from stream-eroded quartzite, white vein quartz, and phyllite that banked against the inner reef margin. Multiple deltas were likely to have formed from drainage courses all along the main axis of the Oscar Range. The Tunnel Creek Road makes the east end of the Oscar Range far more accessible than the west end (fig. 5.2b). But the availability of satellite images and detailed topographic maps also make it possible to explore the remote west end from the comfort of an office.

Contrasting styles of drainage are evident throughout the west end of the Oscar Range (fig. 5.12) based on topographic mapping in an area of 40 square miles (104 km^2). The more common dendritic style is typical of tributaries that coalesce at acute angles.[20] Trellis drainage refers to tributaries that meet at or very close to right angles.[21] This sort occurs in places where folded rock layers result in parallel ridges and troughs with higher elevations capped by harder rocks from which overlying softer rocks have been stripped away, whereas intervening valleys retain the softer rocks. The dominant structure of Paleoproterozoic rocks in the Oscar Range makes a good fit with this scenario in which ridge crests are formed by steeply dipping quartzite beds and narrow valleys preserve the softer phyllite. Territory bordering the longitude of E125° 10' (fig. 5.12) reveals parallel ridges with elevations between 920 feet to 1,050 feet (~280 m to 320 m) above sea level that trace a diagonal pattern trending northwest to southeast. Parallel ridge crests are barely a quarter mile (1.25 km) apart, separated by narrow valleys in which streambeds lead outward to converge on trunk lines cutting across the range at right angles from the northeast to the southwest. Trellis drainage is imprinted within those parts of the range where the ridge crests are highest.

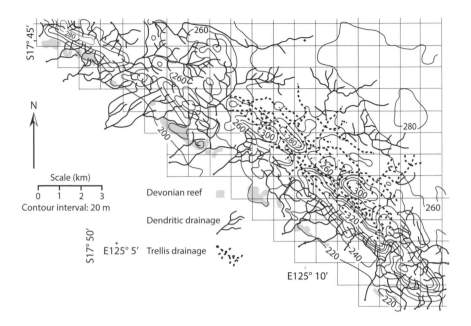

FIGURE 5.12 Topography and stream-flow patterns on the west end of the Oscar Range, showing the difference between dendritic and trellis drainage. See figure 5.2 for outcrop patterns divided between the west and east ends of the Oscar Range.

The folded structure of the Paleoproterozoic Oscar Range is unchanged from 385 million years ago, which is crucial to the notion that Devonian stream drainage may be inferred just as well as observed in today's landscape. The region's post-Devonian history involved burial by Carboniferous and Permian strata, and glaciation left a strong mark on the Australian landscape, particularly in Permian time.[22] Erosion by subglacial meltwater played a role in further sculpting the Oscar Range and parts of the associated great Devonian reef. Initial erosion of the Windjana and Geikie Gorges occurred during Permian time, and the northeast-southwest canyons that cross the Oscar Range were likely enlarged then.

HISTORICAL VIGNETTE: ABORIGINAL SONGLINES ACROSS AN ANCIENT LAND

The average Australian population density is 7.5 persons/square mile (3.2 persons/km²), making the country one of the least crowded in the world. By Western standards, only 10 percent of Australian lands are considered habitable, and as much as 82 percent of the human population is entirely urban. An aboriginal population of 670,000 accounts for only 3 percent of the overall population, which is 25 million.[23] The first inhabitants were the very people who developed the skills necessary to survive a nomadic life in small bands roaming a landscape with limited resources. Since 1976, the Aboriginal Land Rights Act, first enacted in the Northern Territory, has ceded nearly 50 percent of lands in that jurisdiction back to its original inhabitants. Further progress in this movement has also gained momentum in other parts of the country.

Proof of ownership as a basis to assert land rights is a challenge for a people caught in a legalistic, Western-oriented culture, who traditionally led a nomadic lifestyle and lacked a written language. Key aspects of clan life with a strong oral tradition have made a difference in helping to correct the wrongs of dispossession. Central to claims of land ownership is the cultural equivalent of a deeded document in the form of a *tjuringa*, or sacred stone (or carved wooden piece) bearing references to local landmarks. Oral tradition connects each deed to part of a songline that crosses the land to memorialize stories from Dreamtime, when the founders of primordial clan groups were born and roamed the landscape.[24] The principal songlines appear to trace the exploratory paths of the earliest Australians, who entered from the north or northwest from across the Timor Sea.

Clan groups owe parentage to different totem ancestors, such as Bandicoot Man, Cockatoo Man, or Honey Ant Man. The biological siblings of any given family are sure to include individuals belonging to different clans. This is because a totem is assigned to an unborn child at the moment it quickens within a mother's womb. An aboriginal child has no doubt as to the identity of its biological father, but inception by a totem ancestor is linked directly to a physical landmark where the mother was foraging during the early stages of pregnancy and where her child was bonded through a songline with the adventures of a particular Dreamtime ancestor. The stanzas of a song line from a totem ancestor trace the footsteps of that ancestor across the landscape, interacting with a multitude of distinctive rock formations, water holes, steam crossings, and other landmarks. In effect, the entire countryside is interconnected through a web of songlines that anchor the first inhabitants to the land they took possession of thousands of years ago.

While gaining permission from tribal leaders of the Bunaba people to study the rocks around the Oscar Range, our team was advised about the best spots to make camp with access to spring water. Moreover, we were told which particular trees should bear fruit (or not) during our visits according to season. The bond between the land and the people was made clear. Sleeping in a swag on an open cot (as Australian field geologists are accustomed to doing), the Milky Way makes a dramatic appearance each clear night. It is a visible reminder of the aboriginal folklore about the great emu that left footprints as black holes evenly spaced across the bright sands of the southern night sky. The powerful *wand-jina* spirit in charge of the winds that brings the wet season also played a role in our activities. Early rains, unexpected during the last field excursion, flooded our campsite and led to the closure of the Tunnel Creek Road. So it is to experience a landscape with a myriad of stories to tell, including those we impose from an entirely different geological perspective.

LATE DEVONIAN GLOBAL GEOGRAPHY
AND REEF DEVELOPMENT

Reconstruction of Late Devonian (Frasnian) geography includes locations from around the world where major reef complexes developed similar to Western Australia's great Devonian reef (fig. 5.1). Compared to the geography of the Late Silurian (see fig. 4.8), changes in plate tectonics that occurred over the intervening 45 million years entailed the closure of the Proto-Atlantic (Iatpetus) Ocean with the collision of Laurentia and Baltica to form a single continent called Euramerica. The Acadian orogeny led to the rise of Himalayan-style mountains at the junction between the two convergent continents. The master continent of Gondwana underwent a rotation that brought parts of South America and Africa closer to parts of former Laurentia. The narrowed seaway between Euramerica and Gondwana that resulted is called the Rheic Ocean (fig. 5.13). By Frasnian time, the Spanish and French parts of Gondwana were positioned so close to the south flank of Euramerica that through-going circulation to the Rheic Ocean from the east was closed. Another consequence of the Gondwanan rotation brought the Australian segment across the paleoequator to become fully positioned in the Southern Hemisphere. The Panthalassic Ocean was much larger but roughly equivalent in location to the Pacific Ocean. Major circulation gyres powered by the trade winds and westerlies remained largely unchanged from Silurian time in that dominant ocean.

According to the Frasnian map reconstruction (fig. 5.13), the Mowanbini Archipelago and its associated reef complex were situated at a latitude about 17° south of the paleoequator. That the continental Kimberley Block of Western Australia bordered on the sea is solidly established based on extensive Pillara limestone deposits. However, access to the open ocean may have been confined

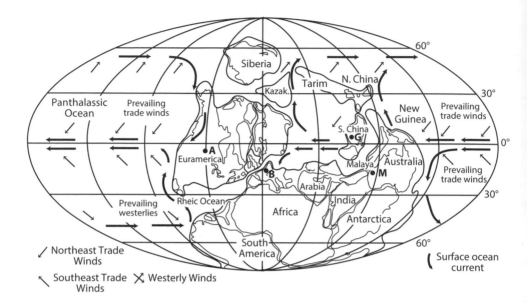

FIGURE 5.13 Global reconstruction for the Late Devonian (Frasnian) world with interpretations for trade winds convergent on the paleoequator, high-latitude westerly winds, and surface ocean circulation. Abbreviations: A = Alberta (Swan Hills Reef Complex) of Euramerica, B = Belgium (European Gondwana), G = Guilin (South China), and M = Mowanbini Archipelago (Australian Gondwana).

to an embayment squeezed between adjacent parts of Gondwanan Malaya and India. The body of marine water in this region opposite the Rheic Ocean is sometimes called the Paleotethys Sea, although the much later Tethys seaway ran between Africa and Europe and between India and Tibet.

Three huge reef complexes from other parts of the Devonian world are similar in size, age, and tropical orientation to the Great Barrier Reef of Western Australia. The Swan Hills reef complex of Alberta in western Canada features many of the same reef and lagoonal stromatoporoids found in the Canning Basin of Australia.[25] The Alberta reef complex is buried entirely belowground and is renowned as a major oil producer. Located about 5° south of the Frasnian equator (fig. 5.13), the structure is nearly central to the Laurentian part of Euramerica and was influenced by gyres of the eastern Panthalassic Ocean. Within the classically studied area around Frasnes in Belgium, another significant reef complex was tenanted by the same key stromatoporoids.[26] Parts of this complex at the margin of European Gondwana are well exposed, but others are known only from the subsurface. The complex is situated at a latitude roughly 15° south of the paleoequator, much like the great Devonian reef of Western Australia. Marine circulation at the west end of the Paleotethys Sea refreshed the Belgium reef complex from the equator, but it is unclear how much of a return current reached the east end of the sea beyond the Turkish and Iranian shores of Gondwana. A third major reef complex is found in the Guilin region of South China, where Frasninan stromatoporoids are characteristic. Platform-margin facies at Qifeng Zhen, for example, are shown to be mounded with as much as 295 feet (90 m) of relief.[27] The landscape around Guilin through which the Li River flows is especially famous for its sugar-loaf topography related to Upper Devonian limestone with prominent reef structures. The Guilin reefs of South China were situated about 5° north of the Frasnian equator (fig. 5.13), separate from Gondwana but geographically much closer to Western Australia than the other reefs.

Extensive shoals and small barrier islands peculiar to reef morphology were certain to have formed in relation to the Devonian reef complexes of Alberta, Belgium, and South China. However, the proximity of those reef tracks to continental rocky shores is nowhere as clear as along the Kimberley Block and Mowanbini Archipelago of the Oscar Range in Western Australia. Exposure of the great Devonian reef in the Kimberleys constitutes a geological feature of highest rank among natural wonders.

SUMMARY

The islands and related spray of islets and sea stacks in the Mowanbini Archipelago rise above a seascape of Devonian Pillara limestone in three-dimensional relief owing to the relative hardness of Paleoproterozoic quartzite ridges, separated as they are by slender valleys of softer phyllite. The inherent stoniness of the countryside captures two realities. On the one hand, there is the contemporary desert with a beauty all its own, with its scattered boab trees and flats dotted by clusters of spiky spinifex grass. On the other hand, there is an ancient coastal terrain beset with a reef track and lagoons on opposite flanks of implicit islands. We are free to wander through both worlds at the same time, and both have stories to tell. The cultural tales refined by its earliest human occupants are not so easy to connect with the geological stories boldly proclaimed in the same landscape. But they coexist side by side.

A distinctly ancient landscape is preserved in stone. It requires hardly any imagination in such a faithfully preserved setting for the geologist to sense the wind and waves in a physical attack against the shores, just as evoked by Rachel Carson in her modern world. Likewise, the geologist is witness to now-extinct forms of marine life that settled into rhythms in a biological response to the same stimuli. Some rocks record daily events with the flux of tides and shifts between onshore and offshore winds. Others testify to the occurrence of episodic storms that washed a deluge of stream sediments out of island hills to bank against the reef complex as tidewater deltas. Yet, there exists a compatible resonance between the Devonian past and cultural present. The *wandjina* wind spirit that regulates the wet and dry seasons of the Kimberleys is deeply imbued in a landscape leading back to the Devonian. The Australian Gondwana of 375 million years ago appears to have generated its own climate, with semimonsoonal seasons arriving through the eastern end of the Paleotethys Sea.

CHAPTER 6

HOW SOFTER ISLANDS DISSOLVE

*A Journey in Early Permian Time to the Labyrinth
Karst of Western Australia*

> *The edge of the sea is a laboratory in which Nature itself is
> conducting experiments in the evolution of life and in the delicate
> balancing of the living creature within a complex system of forces,
> living and non-living.*
>
> —Rachel Carson, *The Edge of the Sea* (1955)

The material composition of rocky shores is as varied as the different kinds of rocks existing under natural conditions. Nature is a great experimenter, and she has the luxury of time to perform her wide-ranging laboratory work. Relative durability is the chief factor under which the shape of a rocky island is influenced over the long run. The archetypal Mount Monadnock gives us quartzite as a metamorphic rock among the strongest and most dense in composition to make entire island groups like the Cambrian Baraboo and Ordovician Jens Munk Archipelagos (see chapters 2 and 3). These proved capable of resisting erosion long enough to be buried in Earth's crust and eventually exhumed in configurations recognizable as former islands. In contrast, water is a short-lived material in its frozen mineral form as ice. An iceberg (i.e., ice mountain) may be considered a floating island that endures as long as it retains its solid form in sufficiently cold water. The effects of global warming were apparent well before the turn of the twenty-first century when great chunks of the Antarctic ice shelf began to crumble and float free from the polar continent. A piece almost as large as the state of Rhode Island broke off from the Larsen Ice Shelf in January 1995 and floated into the Wendell Sea at a latitude 65° south of the equator.[1] The crack in the ice shelf that led to the island's birth was observed to be 40 miles (~65 km) in length and 30 feet (~9 m) wide where the ice shelf was nearly 1,000 feet (~300 m) thick. Who would argue that the physical prominence of such an ice island was comparable, for example, to South Korea's Red island (see chapter 1)? An island remains an island so long as it persists against the assault of erosion.

Between quartzite and ice, nature experiments with the full panoply of solid materials capable of forming a coastline. A combination of different rock types on the perimeter of any given island results in variable topography, prominence, and durability. The story of the Devonian Mowanbini Archipelago in Western Australia (chapter 5) strikes a clear contrast between metamorphic rocks like quartzite and phyllite that make a difference in coastal erosion and patterns of inland stream erosion. In a summary of observed patterns of rocky-shore erosion from around the world,[2] a prominent geomorphologist tabulated a total of 117 studies on sea cliffs formed by sedimentary rocks (51 percent), carbonate rocks (35 percent), soft volcanic rocks like ash and tuff (8 percent), and igneous rocks such as basalt and granite. Among these, volcanic ash and other poorly consolidated ejecta register the highest rates of erosion, up to 260 feet (~80 m) per year. Soft chalk cliffs may retreat as rapidly as 10 to 13 feet (3 to 4 m) per year, whereas hard limestone erodes at rates as low as one inch (2.5 cm) per year. Coastal erosion of sedimentary rocks like sandstone and siltstone is irregular, depending on how well the grains within those rocks are cemented. Normal rates of shore recession for granite and basalt are nearly negligible. Other factors, such as variations in the thickness of sedimentary rock layers or individual basalt flows, also factor into the ability of waves to pry loose whole chunks of rock under prolonged attack. Moreover, how rocks as different as limestone and granite form vertical fractures also facilitate the mechanical work of waves in sculpting rocky shores.

This chapter deals with the formation, burial, and reexposure of labyrinth karst in the form of evenly spaced towers from Lower Permian limestone (Artinskian stage) in Western Australia (see fig. 1.9). In effect, limestone islets laid out in a gridded pattern with canals intersecting at right angles constitute small, loaf-shaped monadnocks with a high vertical prominence in relation to their map footprint. Originally deposited as thick layers that accumulated as marine limestone, dissolution (or karstification) along intersecting joints created passageways that cross as a three-dimensional latticework of caves beneath the surface or as tower blocks rising above a seascape. The historical vignette in this chapter features the maze-like formation of contemporary karst towers in Vietnam's Ha Long Bay, with its floating villages that are home to people earning a livelihood fishing the watery boulevards. In this case, the modern setting of Ha Long Bay offers the best guide to analogous coastal structures from the past.

FEET ON THE GROUND

The coastal town of Carnarvon, at the mouth of the Gascoyne River on the Indian Ocean, is the point of departure for outback lands at the intersection of the Kennedy and Pells ranges within the Carnarvon Basin. There (figs. 6.1a and b), Permian karst topography may be explored in a district centered a little more than 125 miles (~200 km) inland. In a straight line, the Devonian Mowanbini Archipelago is not that far to the present north. But now, with the Devonian mass extinctions of global stromatoporoid reefs behind us, our geological clock is advanced 95 million years for a landing in the early Permian period 280 million years ago.

The goal of a July-August excursion in 1991 was access to the Carnarvon Basin in Western Australia. Literature on the region's geology was reviewed in preparation.[3] Covering roughly 925 square miles (~2,400 km²), a base map (fig. 6.1b) was compiled with an emphasis on Permian strata plotted on several map sheets produced by the Geological Survey of Western Australia.[4] Gascoyne Junction, near the confluence of the Gascoyne and Lyons rivers, has a population of about 150. It is the nearest point of civilization and helpful information on the surrounding district. Outwash plains dominate the landscape between Carnarvon and Gascoyne Junction, sparsely vegetated by mulga scrub, consisting of acacia trees (the thornless *Acacia aneurs*) and tussock grass (spinifex).

We determined that the sheep station at nearby Bidgemia would make the best headquarters for investigating the labyrinth karst as a remnant of a former coastal seascape. The station's owner, Lockley McTaggart, invited us to camp on the banks of the Gascoyne River behind the sheep-shearing shed. Massive eucalyptus trees in the river bed lent a welcome umbrella of shade over the embankment. The crossing at Bidgemia was surprisingly wide, amounting to 0.3 miles (485 m) over the Gascoyne riverbed in the dry season. As the afternoon was young, our truck was put into immediate service on a scouting mission to reach the nearest Permian outcrops. Once on the opposite side of the riverbed, a dirt track led about 4.5 miles (7 km) eastward to the junction with Cream Creek. Another 7.5 miles (12 km) along the creek bed gave access to a pass cut through Permian bedrock (fig. 6.1b). But the exposed rocks showed no signs of karst erosion, and a retreat was made to Bidgemia. The socialization of roosting

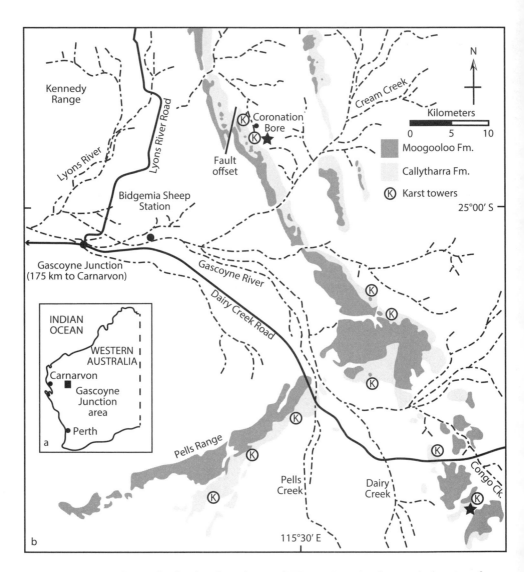

FIGURE 6.1 Geographical and geological maps: a) Western Australia, showing the location of Carnarvon on the Indian Ocean and the inland district around Gascoyne Junction; b) detailed map showing the junction of the southern Kennedy Range with Pells Range and location of several zones with labyrinth karst north and south of Dairy Creek Road. Two locations marked by stars are described in the narrative.

blue-winged kookaburras (*Dacelo leachii*) among the big trees accompanied our preparations for the ensuing evening on the riverbank.

The district covers such a vast area that decisions had to be made as to the most efficient way to canvass the several potential localities for investigation of Permian karst-shore ecology. The next day, a friendly consultation with Mr. McTaggart over the survey maps resulted in an offer of a scouting mission using the station's aircraft, a Piper Cub known for maneuverability at low altitudes. In a sheep drive, the animals are routed from widely dispersed places and chased along a homeward path to the station corral through an action called "mobbing" by a pilot whose task is that of a flying sheepdog working in concert with a small team of riders mounted on motorcycles. There were no fewer than ten locations with notable examples of labyrinth karst marked on the survey maps. A forty-five-minute overflight rendered it possible to choose the best places for on-the-ground inspections. We settled on an afternoon flight schedule when the sun would cast shadows to the east, and I busied myself with laying out a flight plan.

It was a two-seater aircraft, and I was seated close behind McTaggart. The canopy had been removed so that I could shoot film in the open, unobstructed by side windows. The dirt airstrip west of the station compound had a 2,500-foot (0.75-km) runway that was more than ample for the Piper Cub. We were airborne in less than half the distance, taking off toward Gascoyne Junction. The plane banked to the north and then east to follow the Gascoyne River upstream at an altitude of about 300 feet (91 m). The view was breathtaking—a river of trees growing out of a braided streambed with sunlight reflected off small ponds like the sparkle of diamonds. In grasslands with even moderate rainfall, a river with permanent water should support a narrow gallery forest on opposite embankments. In the semidesert landscape dominated by mulga scrub, the gallery forest was embedded as a continuous ribbon within an entrenched streambed. The ecology of the place with water at a high premium was at once apparent.

In no time, the plane reached the previous day's target on our initial foray, where Cream Creek crosses through Permian strata. The plane banked northwest and followed a line of Permian limestone roughly 8 miles (~12 km) to the watering tank at Coronation Bore (fig. 6.1b). McTaggart knew the terrain, and he understood what we sought. Slightly to the east, a splendid example of labyrinth karst appeared below, arrayed like so many apartment blocks on a regular town grid. Shadows cast by the afternoon sun perfectly outlined the galleries that separated row after row of karst towers eroded from the Callytharra Formation (fig. 6.2). A small butte sat like a cap over a portion of this karst landscape. It was formed by superincumbent Permian sandstone belonging to the Moogooloo

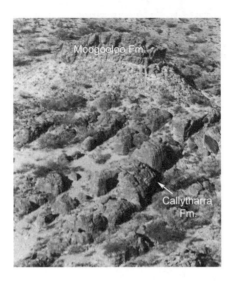

FIGURE 6.2 Aerial view of labyrinth karst near Coronation Bore as developed in limestone belonging to the Permian Callytharra Formation (foreground) with an outlier of sandstone belonging to the succeeding Moogooloo Formation (background). Coordinates: 115° 25' 52" E, 24° 54" 43" S.

Photo by author.

Formation, which concealed not even a quarter of the labyrinth. McTaggart turned to look back at me, and I showed my excitement with repeated gestures of an arm pumping a thumb's up signal. Circling the butte, McTaggart brought the aircraft closer to the ground, and I shot image after image as fast as possible. This was surely it—a superior locality we must revisit on the ground.

The plane headed south, crossing the Gascoyne River and the adjacent road to reach Permian strata in the Pells Range (fig. 6.1b). Along the east side of the range, we followed a succession of karst zones, none of which surpassed what we saw near the Coronation Bore. Yet another zone of karst erosion in the Callytharra Formation was spotted as we completed a figure-eight loop around the fringe of massive Moogooloo sandstone. Onward to the southeast, our route took us back over the Gascoyne River and crossed the roadway to survey a smaller exposure of sandstone overlying the limestone. South of Congo Creek (fig. 6.1b), a fine example of labyrinth karst appeared with ranks of towers not as high as those near the Coronation Bore. Two localities with excellent karst architecture out of ten possible candidates were identified, and our plan for the next few days became clear.

In a straight march, the shortest distance from Bidgemia Station to the Coronation Bore amounts to 11.5 miles (~19 km). Morning dawned with a soft breeze that promised to make the day's excursion more comfortable. Water bottles were filled and lunches packed, and we retraced our earlier path by truck to the area where Cream Creek crosses Permian strata as a tributary to the Gascoyne River. This brought us within reach of the Coronation Bore on an outbound hike of about 6 miles (~10 km). The northerly trek leads across ground that affords comparisons between the Callytharra and Moogooloo Formations. The Moogooloo Formation was a fine- to medium-grained sandstone dominated by quartz grains. Scarce trace fossils suggested a marine influence on the sandy beds closest to the contact with the underlying Callytharra Formation. Higher still, silty intervals separated thicker layers of sandstone distinguished by trough cross-bedding. Not unlike present-day stream activity under flood conditions on the Gascoyne River, these structures indicate that fluvial processes were active during Permian time.

In contrast, exposures of the underlying Callytharra Formation showed thin- to medium-bedded limestone layers formed by abundant fossil fragments. Overall, the contributing fauna was reduced to sand-size bits of calcareous material (a calcarenite, in more technical language). Here and there, larger fossil pieces were represented by various marine invertebrates, including brachiopods, mollusks, shelled squids, corals, and bryozoans. Indeed, the Callytharra Formation is regarded as one of the most fossil-rich units in the Carnarvon Basin.[5] The stratigraphic juxtaposition of the two formations could not have been more telling of a profound change in conditions. Even so, the development of an erosional surface at the interface between the two formations points to an interval of subaerial emergence during which little or nothing was deposited as the dissolution of the Callytharra limestone began to form karst towers prior to the arrival of the Moogooloo sands.

At ground level, arrival at the Coronation Bore locality yielded a sight nothing short of dramatic. The cap rock on the butte over the labyrinth karst (fig. 6.2) is estimated to be about 150 feet (45 m) in diameter, but with descending slopes of softer rock that triple the radius of the Moogooloo outlier. From these slopes, the steep side walls of the Callytharra karst emerged like partially excavated structures belonging to some kind of Maya temple complex, where the supervising archeologist called a sudden halt to fieldwork and abandoned the dig. Nature enacted the complete burial of this geological site, and nature was still uncovering it. Clean lines that traced the horizontal layering of limestone on vertical walls drifted against and faded away under a cover of engulfing sandstone (fig. 6.3a). In a few places, some sandstone still adhered to the sides of the darker karst blocks. They appeared like splotches of flaking paint yet to be removed by a good power wash.

The larger structure drew the visitor into a network of narrow passages and broader corridors. Side alleys were found to be from 6.5 to 10 feet (2 to 3 m) wide, separating block towers that rose steeply 20 to 26 feet (6 to 8 m) in height (fig. 6.3b). Relative prominence could be greater than it appeared, as flooring on the byways was not entirely excavated from the soft sandstone. The primary corridors were between 33 and 65 feet (10 and 20 m) wide. Exploration quickly instilled a feeling of disorientation, even though the central, castle-like presence of the sandstone butte was rarely out of view. Minutes stretched into hours devoted to a search for signs of organic borings or encrustations on the walls of the karst towers. It was disappointing that much of the vertical space was covered by sharply fluted surfaces, reactivated by carbonate dissolution after exhumation from the sandstone cover. The labyrinth walls were not in pristine condition. They had been so etched by time that the "mural paintings" of former sea life we might have hoped to find preserved in place were erased. As the sun sank ever lower on the western horizon, it was with regret that we started the long hike back. Our last water consumed; the day exacted a toll with dehydration despite

FIGURE 6.3 Ground views showing details of Permian strata near the Coronation Bore (see fig. 6.1 for map location): a) wall of Callytharra limestone protruding from under the cover of overlying Moogooloo sandstone; b) karst towers eroded from bedded limestone arranged along regular corridors (person for scale).

Photos by author.

cooler than anticipated temperatures. Back at Bidgemia for the night, the notion of a cooked meal was unappealing. Instead, we raided our larder of canned fruit (peaches and pears packed in thick syrup) to satisfy our thirst.

STRATIGRAPHIC POSITION AND RELATED GLACIAL FEATURES

To better understand the layered context of the Callytharra and Moogoolo Formations, it is necessary to consider other rock units that preceded and followed after in geologic time (fig. 6.4). The Callytharra is one of the thickest rock units in the Carnarvon Basin, amounting to as much as 870 feet (265 m). Thickness generally relates to the amount of time over which rocks are deposited. However, in certain cases, very thick deposits may accumulate extremely fast, as, for example, due to submarine landslides. Limestone, however, does not easily fit that category, and it may be assumed that the Callytharra Formation took a long time to accumulate. Less than a third of the thickness of the Callytharra Formation, the preceding Carandibby Formation is formed mainly by interbedded shale and siltstone. It includes some fine-grained limestone higher in the sequence. Most significant is the presence of glacial erratics within the lime-rich strata (fig. 6.4). A geological erratic is represented by rocks that normally have no business being associated with the background lithology in which they occur. In this case, glacial erratics in the Carandibby Formation consist of pebbles and small cobbles of granite carried off a mainland by advancing ice flows and further transported seaward on ice floes. Such erratics also are called dropstones because, once the ice melts, its load of glacially scoured material sinks to the bottom of the sea.

Succeeding the dominant sandstone of the Moogooloo Formation is the Billidee Formation, which, like the Carandibby Formation long before, marks the reappearance of lime-rich strata with glacial erratics. After the Billidee Formation comes the Keogh Formation, which is similar to the Moogooloo in that it is dominated by clastic rocks like siltstone, sandstone, and conglomerate (fig. 6.4). Hence, much of the Carnarvon Basin preserves a pattern of repetitious clastic rocks with longer or shorter intervals filled by limestone. Limestone is most typically associated with warmer climates (although cold-water carbonates occur in places like New Zealand today). The unmistakable imprint of Permian deposition in the Carnarvon Basin is that the contrasting lithologies of the Callytharra and Moogooloo Formations are bracketed by rock units

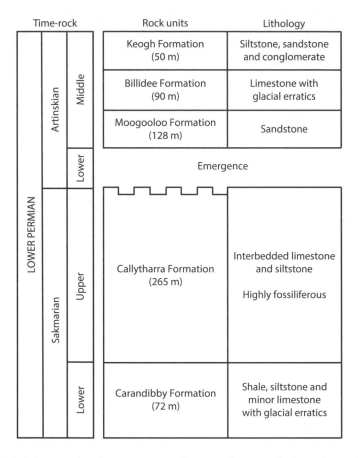

FIGURE 6.4 Stratigraphic chart summarizing the major depositional relationships and timing of Lower Permian strata in the Gascoyne Junction region. The labyrinth karst is eroded in the upper part of the Callytharra Formation (indicated by the notched line).

containing glacial dropstones. Fluctuations in climate occurred through time, but the old river courses and sea beds preserved in the Carnarvon Basin were never far away from glaciated parts of Australia within the larger geographic context of Pangaea.

During what part of the Permian period was the labyrinth karst at Gascoyne Junction under active development? The Permian period lasted for about 48 million years (see fig. 1.9), and all available evidence suggests that the karst towers sat exposed in a coastal zone during the early Artinskian roughly 280 million years ago before burial by Moogooloo sands some fifteen or twenty million years into the Permian period. Karst topography could not begin to be shaped until after the Callytharra Formation was duly deposited as marine limestone as part of the antecedent Sakmarian stage, and the best fossil evidence for the timing of that unit puts limitations on the subsequent dissolution of joints. Professors at the University of Iowa, with whom I studied as an undergraduate geology major from 1967 to 1971, were foremost in researching the biostratigraphy of Permian ammonites worldwide.[6] Their work established the relative ages of these local units based on evolving shell structures in coiled cephalopods (squids). Formally, the top of the Sakmarian stage is defined by key fossils at the base of the succeeding Artinskian stage based on the first appearance of key conodont species.[7] For now, the Permian strata of Western Australia are correlated with respect to master sections in Russia's Ural Mountains. Local relationships that stand out most clearly demonstrate how the Callytharra and Moogooloo Formations were separated in time by an interval of subaerial emergence. Moreover, the Carnarvon Basin experienced glacial ice flows before the Callytharra limestone was deposited and after the Moogooloo sandstone entombed the labyrinth karst.

FOSSIL EVIDENCE ON PERMIAN KARST?

The site southwest of Congo Creek, scouted by air (see fig. 6.1b), proved more exciting in the hunt for encrusting or boring fossils associated with the three-dimensional erosion surface on the Callytharra karst. Although the karst towers at that locality were not as high or as extensive as explored near the Coronation Bore, previously unreported features proved interesting. Foremost was the discovery of a silicified crust, approximately one-eighth of an inch (5 mm) thick, that formerly covered the walls of the karst towers (fig. 6.5a). Only patches of the original crust remain intact, the largest with an area of 125 square inches (~800 cm^2). The crust itself is locally coated with a red-to-black rock varnish. A closer view of one such patch shows how it sits on a pedestal of underlying limestone, and the surviving wall face is deflated by an inch or more (2 to 3 cm) of erosion. Sharp flutes on the deflated surface (fig. 6.5b) demonstrate the reactivation of karst features during more recent times. Other karst features on the tops of some towers correspond to small solution basins (kamenitzas). Later laboratory work based on energy-dispersive x-ray spectroscopy (EDAX analysis) focused on the mineral composition of samples from the crust. A bulk sample gave a value of nearly 80 percent silica, whereas a more constrained sample yielded a value > 90 percent for lamellae preserved within the crust. The surface

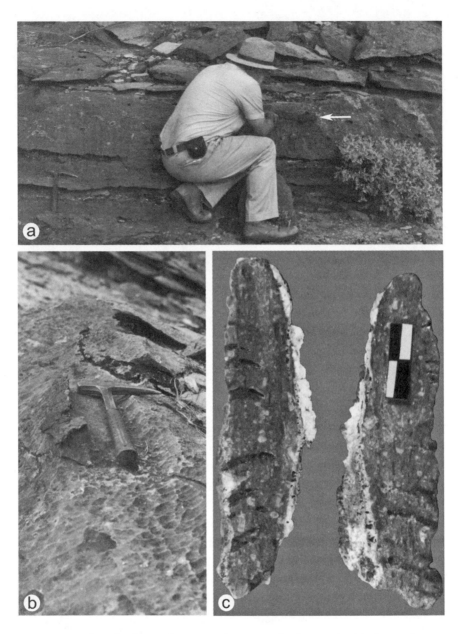

FIGURE 6.5 Details of tower karst near Congo Creek: a) vertical wall of tower karst in Callytharra limestone showing remnant patches of siliceous crust in two places (hammer handle in the lower left, marks one of them and white arrow in the middle right points to another); b) close-up view of same siliceous crust marked by white arrow; c) cut and polished layer of limestone from the top of a karst tower, showing subparallel tubes penetrating below the surface (scale = 6/8 inch or 2 cm).

Photos by B. Gudveig Baarli (a) and the author (b, c).

beneath the crust consisted of 75 percent calcium-carbonate ($CaCO_3$) and close to 15 percent silica (SiO_2). The silica-rich crust was likely derived from local dissolution and reprecipitation of silica originally deposited in the Callytharra Formation. As such, the covering provides more evidence of chemical alteration of the original tower walls.

Organic structures preserved on top of the karst towers as subparallel tubes also were found to extend as much as 5/16 of an inch (15 mm) into the limestone (fig. 6.5c). Specialists consulted on the possible identification and relevance of these structures could not agree on their origin or their true age. The more conventional interpretation is that the tubes were borings produced through bioerosion by Permian polychaetes (marine worms). A dissenting opinion favored an interpretation involving calcium-carbonate absorption and excavation by the rootlets of modern land plants. The dispute remains unresolved, and further fieldwork is necessary to determine if the tubes might be excavated from beneath a cover of Moogooloo sandstone (i.e., from the top of karst towers still partially buried, as shown in fig. 6.3).

HISTORICAL VIGNETTE:
KAYAKING IN VIETNAM'S HA LONG BAY

Ha Long Bay sits within the Gulf of Tonkin, occupying a core zone of some 168 square miles (435 km²) on the northern coast of Vietnam. The region is legendary for its drowned labyrinth karst featuring a maze of some 1,600 islets that rise as limestone towers in a sheltered seascape. In 1994, the district was added to the UNESCO World Heritage List. Additional protections derive from Vietnam's designation of this unique area as a Special National Landscape Site under the Cultural Heritage Law of 2009. Located some 110 miles (~180 km) east of Hanoi, tourists throng to the zone for its scenic beauty, accessible by tour boats. The name, which translates as place of "descending dragons," is fixed in national mythology but appeared on regional maps only in the later 1800s. A mother dragon and her host of fledging dragons are credited with creating a dense-island bulwark against invasion by external enemies. Some of the larger islands showcase karst caverns that inspire fables and other stories all their own.

For the geologist and those with a passion for the artistry of landforms, the labyrinth karst is the prime attraction for exploration by kayak. Regarding stratigraphy, Ha Long Bay shares certain aspects with Australia's Carnarvon Basin. The limestone formations that contribute to Vietnam's karst structures were laid down during middle Carboniferous to early Permian times. Limestone from the Cát Ba Formation is Carboniferous in origin and as much as 1,475 feet (450 m) in total thickness. The Quang Hanh Formation is an even thicker amalgamation of Middle Carboniferous to Lower Permian limestone, amounting to 2,450 feet (750 m). Whereas karst topography developed during Permian time in Western Australia, the Ha Long seascape, with its similar labyrinth karst, waited to develop only during the last 2,000 years. The same processes took place, however, with dissolution along regularly spaced joints in the limestone that slowly widened into galleries both exposed at the surface and below ground.

The immediate sensation for the kayaker on a tour expected to last more than a couple hours is the realization that few places exist where the kayak can be beached, and the occupant may pause to stretch her legs on dry land. Vertical cliffs around each islet plunge straight into the water with no rock shelf in sight. Overall, the water depth within any given corridor rarely exceeds 33 feet (10 m), although the wider channels are much deeper. Disorientation is another reaction. Tour leaders advise that it is hard to get lost if the kayaker consistently turns to the right (or consistently to the left) at intersections. Once these negative sensitivities abate, the sheer beauty of the place takes over, especially in exploration of the side galleries. On a clear day, the green of the foliage on the steep cliffs could not be any greener, and the splendor of plants like the endemic climbing shrub (*Jasminum alongense*) or endemic palm (*Livistoma halongensis*) exert a calming mood. The density of karst islets is great, and individual towers commonly exceed 165 feet (50 m) in height (fig. 6.6a).

On average, the tidal range in Ha Long Bay is nearly 10 feet (3 m). One of the advantages of a kayak excursion is its proximity for close examination of the sea cliffs concerning the intertidal zone. A common geomorphic feature around many islets is a tidal notch (fig. 6.6b), visible during low tide as a distinct mark. Particularly around the outer line of islets where wave exposure is greatest from the open sea, cliffs are subject to collapse due to being undermined. In the tidal zone, the rock oyster (*Crassostrea* sp.) is ubiquitous and densely populated. The dotted periwinkle (*Littorina pintado*) also makes an appearance. Marine invertebrates of this kind did not evolve until well after the Permian period.

The natural history of Ha Long Bay includes its human population of fisher folk, whose history may be traced back thousands of years. As there is no land to build a village, floating villages developed a culture distinct from mainland Vietnam. Fewer villages exist now because the government has begun to move people to the mainland, where regular schooling is available to children. Some villages continue to thrive, like Dong Câm, near a spectacular sea arch accessible

FIGURE 6.6 Tower karst in Ha Long Bay, northern Vietnam: a) far view showing multiple towers in a maze-like setting; b) closer view of the central tower showing a distinct tidal notch (white arrows).

Photos by author.

at low tide by small boats with a typical flat-bottom design. Village women, for the most part, earn an income from tourists who board their modified craft two at a time to be rowed out through the sea arch (fig. 6.7). With 5.5 million tourists visiting Ha Long Bay annually, however, there are worries about over-exploitation of the environment and degradation due to water pollution.

FIGURE 6.7 Village woman from Dong Câm prepared to receive visiting tourists to a nearby sea arch eroded in Permian limestone.

EARLY PERMIAN GLOBAL GEOGRAPHY

The assembly of supercontinent Pangaea began to take shape during the Early Permian (fig. 6.8). The South China Block and parts of Indo-China were the largest fragments not yet joined to the master continent during Artinskian time 280 million years ago. Compared to the earlier global array of continents for the Late Devonian (see fig. 5.10), the Rheic Ocean became fully closed off, and Euramerica joined with neighboring parts of Gondwana on the opposite side of the Rheic Ocean in present-day South America, Florida, Africa, and southern Europe (Spain and France). Present-day Siberia and Kazakhstan were linked together with western Russia and Scandinavia. The Panthalassic Ocean remained the dominant body of water on the planet, with strong circulation gyres in the Northern and Southern Hemispheres. As in the Devonian, however, the Paleo-tethys Sea continued as a significant feature with small circulation gyres north and south of the paleoequator.

As it relates to the Australian segment of Pangaea and the Carnarvon Basin, a peripheral current off the southern gyre in the Paleotethys Sea brought normal seawater to the shelf of Western Australia during deposition of the fossiliferous Callytharra limestone at the close of Sakmarian time. At the transition between Sakmarian and Artinskian times, Gascoyne Junction was located at a high latitude around 65° south of the paleoequator (fig. 6.8). The question remains whether or not circulation was sufficient to bathe the labyrinth karst in normal seawater as the sands of the Moogooloo Formation began to pour into the district. Marine invertebrates in the *Eurydesma* fauna are recorded from widespread localities in eastern Australia in rocks roughly equivalent in age to the Callytharra Formation.[8] *Eurydesma* was a large bivalve (clam) adapted to a rough-water environment in cobble beds adjacent to rocky shores. That part of Australian Pangaea was positioned at an equally high latitude, sure to have been refreshed by the southern gyre of the Panthalasic Ocean (fig. 6.8). Cold ocean water apparently was not a detriment to the *Eurydesma* fauna. The paucity of marine fossils preserved as encrusting organisms or boring traces on the karst towers of Gascoyne Junction remains a conundrum, as does the absence of tidal notches like those in Vietnam's Ha Long Bay.

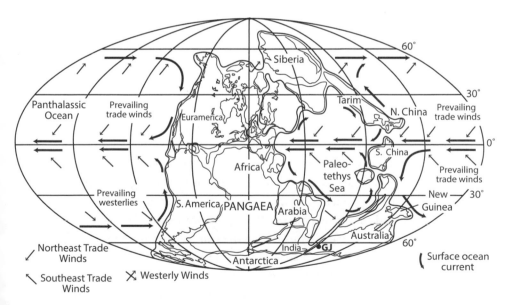

FIGURE 6.8 Global reconstruction for the Early Permian (Artinskian) world with interpretations for trade winds convergent on the paleoequator, high-latitude westerly winds, and surface ocean circulation. GJ = Gascoyne Junction area in Western Australia.

Discharge of glacially-fed waters carrying Moogooloo sands to the shores of Pangaea's Western Australia may have mixed with and diluted normal seawater to the extent that diverse marine life was prohibited from the galleries of the labyrinth karst. If the true height of the karst towers is not fully exposed, the possibility remains that Permian tidal notches are yet to be discovered. In other words, the effect of the earliest rise in sea level after subaerial exposure and erosion of the Callytharra limestone may yet be hidden from sight under the cover of Moogooloo sands. The burrows of marine organisms preserved in the lower part of the Moogooloo Formation imply that the underlying karst topography was part of a former coastal setting, at least for a time. The glacial dropstones in strata preceding and following the Callytharra-Moogooloo sequence remind us that the Permian shelf of Western Australia was at a latitude comparable to the present-day Larsen Ice Shelf on the Wendell Sea of Antarctica.

CHALLENGE OF PANGAEA: THE LARGEST ISLAND

Pangaea as a supercontinent was complete by the end of the Permian period 250 million years ago. It persisted as the world's largest island well into the following Triassic period until the next 12 million years, when the North Sea between the present-day British Isles and Scandinavia began to open up. The Triassic system is a challenge to the search for well-defined paleoshores, as categorized by published studies on rocky shores, sandy shores, and muddy shores.[9] Out of 150 studies canvassed on geological unconformities interpreted as rocky shores, none relate to the Triassic system anywhere in the world. Due to convergent plate tectonics that led to Pangaea, the preponderance of passive shorelines in most places around the mega-island meant that rocky shores were less likely to be developed. The transition between the Permian and Triassic periods also appears to have been an interval of lower global sea level, during which great river systems draining the supercontinent were likely to transfer massive amounts of sand and mud across broad continental planes to the global shoreline. That transition 250 million years ago also happens to record the third and most devastating mass extinction to affect the blue planet. Smaller Triassic islands with a distinctive fauna of pecten bivalves (*Monotis* sp.) are known to have existed, but they all became accreted to the supercontinent as suspect terranes lacking distinctive rocky shores.[10] The original location of these islands in the Panthalassic Ocean cannot be traced.

SUMMARY

Layers of thick limestone in the Callytharra Formation were deposited prior to 280 million years ago at a high latitude but during a relatively warm interlude between recurring glaciations in the Australian sector of Pangaea. The limestone's rich fossil content is proof positive that the Carnarvon Basin was flooded by normal seawater. Surface exposure of the limestone during a subsequent drop in the sea level led to dissolution along vertical joints that gradually deepened into a three-dimensional karst labyrinth. The burial of block-shaped karst towers by sands in the Moogooloo Formation followed in Artinskian time when glacial meltwater fed a network of interconnected (braided) streams carrying sediments across a wide coastal plain. Trace fossils left by organisms that burrowed into soft sediments are preserved in the lower part of the Moogooloo Formation, but ocean water may have been diluted as brackish water due to the high volume of freshwater discharged.

Evidence for marine invertebrates that may have encrusted or bored into the side walls of the karst towers is notably absent. Borings at the tops of some karst towers can be authenticated as Permian in origin only if excavated from beneath the cover of the Permian Moogooloo Formation. When the dark crust lining the walls of the karst towers actually formed is unclear, but silica chemically remobilized from the underlying limestone is more likely than a biological source at the surface. The ongoing development of small-scale dissolution features during recent times is clear from the deflation of the limestone surface where the silica crust is worn away. The beautiful but austere lands of the Carnarvon Basin have more geological secrets to disclose regarding nature's experimental laboratory. Rachael Carson's observations on the subject are prescient in this regard. Comparison with the modern labyrinth karst of Vietnam's Ha Long Bay further informs us that coastal karst towers may result under both tropical and more polar conditions so long as abundant surface water is freely available to dissolve limestone.

CHAPTER 7

HOW ISLANDS REACT TO BIG STORMS

A Journey in Early Jurassic Time to
Saint David's Archipelago of Wales

As long as there has been an earth, the moving masses of air that we
call winds have swept back and forth across its surface. And as long
as there has been an ocean, its waters have stirred to the passage of
the winds.

—Rachel Carson, *The Sea Around Us* (1951)

Ecological and geologic time operate from different but intertwined parameters regarding rock making. The former is all about the moment, like stop-action frames from a movie picked at random and set in stone from a continuous loop played out by physical and biological processes. The latter is more typical of the rock record as an amalgamation of disconnected moments compressed into a blurry image. The challenge for the paleoecologist is to sift through the rocks in search of rare ecological events that most clearly depict the interaction between the wholeness of life and the physical conditions that constrain it. The rock record is further complicated by time gaps or holes in the registry, like the time between the dissolution of a karst topography and its burial (see chapter 6). Charles Darwin compared the rock record to a fragmentary book with every other chapter ripped out and only a few lines preserved from each page.[1] Discovering those ecological moments preserved in rock is akin to deciphering paragraphs or even a whole page from a very old and one-of-a-kind book.

Any given island at any given place will experience a multitude of good days as well as some bad days when life is in peril. To those who spend even an entire lifetime at the shore, the winds that stir ocean waves seem fickle in their ever-changing harmonies. Days on end are quiet except for the rise and fall of the tides. Some seasons bring more storms than others. Certain hurricane seasons are remembered as more violent than others. If nothing else, humans aspire to measure and categorize things related to natural phenomena. In 1805, British naval officer Francis Beaufort devised a scale from zero to twelve with standardized observations linked to the behavior of wind-driven waves on the open sea.[2] Calm is denoted by a zero when the ambient wind speed is < 1 mph (0.3 m/s). A six is for a strong breeze from 25 to 31 mph (10.8 to 13.9 m/s), during which spray is blown off the tops of large waves up to 9.9 feet (3 m) high. A gale under wind speeds between 55 and 63 mph (24.5 to 28.4 m/s) and accompanied by waves up to 29.5 feet (9 m) high merits a 10. On the Beaufort scale, a 12 is for hurricane-force winds ≥ 73 mph (≥ 32.6 m/s). The Saffir-Simpson scale takes over from the Beaufort scale with five categories for hurricanes that start with wind speeds of 74 mph (33 m/s) and reach the highest ranking of a Category 5 hurricane starting with wind speeds of 157 m/h (70 m/s). Island shores and island life are vulnerable to direct hits from big storms and hurricanes, and the rock record tells some of that story from deep time.

This chapter examines Lower Jurassic rocks around a paleoisland in Saint David's Archipelago built on a foundation of Carboniferous rocks (also Triassic rocks to a lesser degree) in the Vale of Glamorgan in South Wales (UK). Parts of an island rocky shore are exposed on the north coast of the present-day Bristol Channel, distinguished by conglomerate from the Jurassic Sutton stone resting on thick-bedded Carboniferous limestone. The time gap between these two principal rock types is about 145 million years. Details regarding ecologic time are subtle because they feature Jurassic organisms that secreted carbonate colonies and shells attached directly to the Carboniferous limestone and cobbles eroded from that older limestone. The Sutton stone conglomerate is controversial on its own due to disagreements over the severity and frequency of storms that battered the Jurassic shores of Saint David's Archipelago. From a historical outlook, the unconformity in this district entails the longest period of studies on any former rocky coastline currently known to science, dating back to 1846. The vignette on human relationships to the land in this chapter looks at the contributions of amateur paleontologists and the rise of a truly professional class of geologists.

FEET ON THE GROUND

The Welsh town of Bridgend is the gateway to a stretch of coastline 4.5 miles (7.25 km) south on the Bristol Channel between the village of Ogmore-by-Sea and Dunraven Bay. In turn, Bridgend is located equidistance 20 miles (32 km) west of Cardiff and east of Swansea. Generations of students from those Welsh cities have visited Ogmore-by-Sea to walk the coast and study the local geology under the supervision of their professors. My summer excursions to Ogmore-by-Sea were undertaken in the early 1990s in the company of an Oxford mentor who shared an interest in paleoshores and agreed to cosponsor a formal study of the area published a few years later.[3]

The Bristol Channel separates Wales in the north from the English Mendip Hills of Somerset to the south, with the Severn River entering the channel from the east (fig. 7.1a). One of the largest wetlands in the British Isles, the Severn Estuary features the second-largest tidal range in the world, amounting to 40 feet (~12 m). Nova Scotia's Bay of Fundy holds that record with a tidal range of 53 feet (16 m). My good fortune has taken me to both places. Nothing quite compares with the enormity of open ground when the tide is fully out, and nothing is quite so unsettling as the rapidity with which the tide returns across the seabed. In planning a day trip to Ogmore-by-Sea, checking the tide table in advance is paramount. The daily high tides are separated by twelve hours and twenty minutes. If photography is planned, it is crucial to have the morning sun strike the shore to catch the best light against the Carboniferous limestone and overlying Jurassic Sutton stone in sea cliffs facing to the southwest.

The weather and timing for today's visit are ideal, with low tide predicted at 10 A.M. It is 9:00 when we leave the vehicle at the car park south of Sutton (fig. 7.1). Less than a mile's (1.5-km) walk is ahead, starting off well above mean sea level on the way to Pant y Slade (Welsh for "slade's hollow"). The coastal path on the protected Heritage Coast crosses an open landscape that is windswept and dominated by a grassy ecosystem including false brome (*Brachypodium sulvaticum*), glaucous sedge (*Carex flacca*), and crested hair-grass (*Koeleria macrantha*).[4] Plants more tolerant of sea spray, like the red fescue (*Festuca rubra*) and flowering sea carrot (*Daucus carota*), are more prevalent on the slopes below. Where exposed, Triassic rocks belonging to the Mercia mudstone are cut by veins of calcite that include pieces of lead ore (galena). A reddish hue stains the Jurassic rocks in contact with the Triassic. In turn, those Triassic rocks fill channels carved into the underlying Carboniferous limestone.

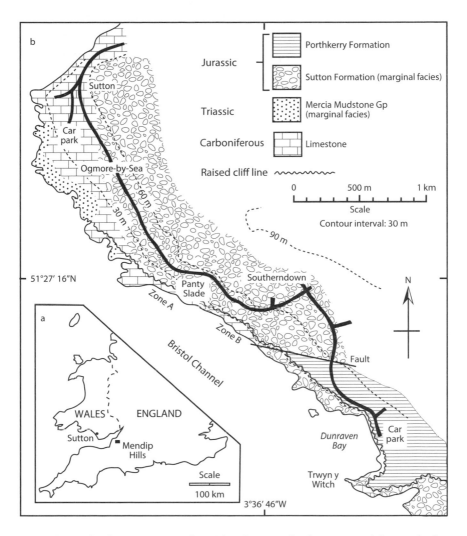

FIGURE 7.1 Combination topographic and geologic map for the area around Ogmore-by-Sea on the Bristol Channel, South Wales (after Johnson and McKerrow, 1995, their fig. 1).

At Pant y Slade, we first encounter the Sutton stone where it sits on thick-bedded Carboniferous limestone. Our geological clock is now reset at 198 million years ago. The lapse in time since leaving the Permian karst islets of Western Australia is almost 100 million years. With the ease of a champion pole vaulter, not only have we jumped over the end-Permian mass extinction believed to have extinguished species belonging to 80 percent of marine invertebrate genera but also the end-Triassic mass extinction that exterminated additional marine species as well as many land-dwelling animals, among which were mammal-like reptiles adapted to life with early characteristics of warm-bloodedness.[5] Evidence points to extreme climate change perturbed by massive continental outflows of flood basalts, both at the close of the Permian period, marking the end of the Paleozoic era, and the close of the Triassic period in the Mesozoic era.[6]

The contact surface between Carboniferous and Jurassic limestone where we now set foot is part of a wave-cut platform, flooded today only by storm waves. Due to the striking difference between Jurassic conglomerate and Triassic limestone, the junction is distinct. Both kinds of strata are crudely horizontal in relationship to each other, making recognition of an unconformity more challenging. At zone A on the rock platform (fig. 7.1b), a subtle set of swales or shallow channels is apparent in the Carboniferous limestone (fig. 7.2). From the center of one trough to the next, the width is more-or-less 40 inches (1 m), and the depth below adjacent crests amounts to no more than a foot (30 cm). The swales are parallel to one another and disappear beneath the overlying Sutton stone. Here, the Jurassic consists of reworked Carboniferous pebbles and cobbles, somewhat worn but not in direct contact with one another, embedded in lime sand.[7] Showing no signs of internal bedding, the conglomerate buries the swales under a thickness of about 70 inches (1.75 m). The observer might be persuaded that the Sutton stone arrived all at once, dumped as a single package. Quite a different kind of conglomerate would be expected to materialize from beach gravel, where all the individual stones are in direct contact with one another. In such a case, the energy of the waves against the shore can be relied on to wash away the fine detritus that results from clasts rubbing against one another in constant agitation. Hence, it may be concluded with some confidence that we do not find ourselves standing on a former beach.

In our eagerness, however, we are getting ahead of the story. Retrieving rubber knee pads from our day packs, it is time to get down to business. Aside from the topography of the undulating swales, the surface of the Carboniferous limestone appears to be clean. On hands and knees, the goal is to search the surface for evidence of Jurassic organisms attached to or bored into the Carboniferous rocks. The search area available for scrutiny is narrow but extends laterally for 425 feet (130 m). In effect, we have a hard-rock seabed that was swept by long-past Jurassic tides to erode the shallow swales at our disposal. Eyes trained close to the stony surface, the task seems fruitless at first. But the weather is calm, and the present-day tides of the Bristol Channel won't return anytime soon.

They are easy to miss, but low domed-shaped colonies belonging to Jurassic corals are found cemented in growth position on the Carboniferous rock surface. They look more like fat pancakes only 6 inches (15 cm) in diameter but are

FIGURE 7.2 Tidal rills eroded at the surface of Carboniferous limestone overlain by conglomerate belonging to the Jurassic Sutton stone (from Johnson and McKerrow, 1995).

Photo of the late W. S. McKerrow by author.

well-worn by exposure to the present-day elements of rain and sea splash. Faint signs revealing the hexagonal outline of calyxes where individual coral polyps sat in place are barely 2/5 of an inch (1 cm) in diameter but retain traces of the septa that lined each cup with radiating partitions (fig. 7.3a). Compared to the encrusting Ordovician corals from Hudson Bay (chapter 3), descendants of which went extinct at the end of the Permian period, these are advanced scleractinian corals (*Heterastraea* sp.) related to modern stony corals. The spatial density of the colonies is low, although it cannot be excluded that some have been worn away. What is clear, however, is that these corals preferred to expand outward rather than upward to any significant amount. In form, they are solidly constructed, meaning that the cups holding neighboring polyps are pressed together with no open space between them. The hexagonal packing of the calyxes in growth is a result. Another coral species with much smaller, star-shaped calyxes (*Allocoeniopisis gibbosa*) also occurs fastened to the Carboniferous limestone surface. This is exacting work, as it is necessary to crawl on all four limbs with a hand lens ready to examine suspect fossils under magnification.

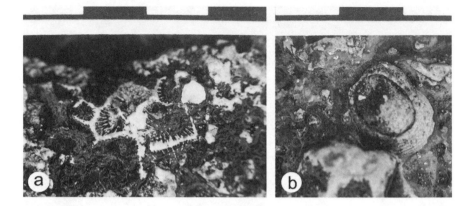

FIGURE 7.3 Jurassic fossils on Carboniferous limestone: a) hexagonal-shaped calyxes from a coral colony (*Heterastraea* sp.); b) dorsal view of an elbow shell (*Patella suttonensis*). Centimeter scales.

Photos by author.

Other fossils that catch the eye are oysters (*Liostrea hisingeri*) with only one valve left in place attached to the rock surface. These bivalves first appeared in the early Jurassic, leading to a successful line that continued to proliferate right down to the present day.[8] A small elbow shell or limpet (*Patella suttonensis*) also is in evidence (fig. 7.3b), representing a Jurassic mollusk that clung to the rock surface by the suction of its fleshy foot. Yet other traces appear as large-diameter and small-diameter borings in the limestone. The larger holes were made by torpedo-shaped clams that secreted mild hydrochloric acid to etch out dwelling places below the surface of the limestone. The opening at the surface is where the clam's siphons were positioned to bring nutrients and dissolved oxygen to the animal hidden below. Typically, only the vacant space where the clam was imprisoned is left as a trace fossil (*Gastrochaenolites* isp.). The much smaller holes are the openings to straw-shaped borings where sea worms made their home. Here, the trace fossils (*Trypanites* isp.) are mainly vertical in orientation and may be up to 3 inches (7 cm) in depth, viewed in cross-section. These marine polychaetes may have looked like the tube-dwelling "Christmas tree" worm that extrudes its delicate feeding apparatus into the water to collect microscopic food items but quickly withdraws it when threatened from above. Time passes rapidly in our pursuit of the rare ecological moment captured by fossil remains at the surface of the rocky seabed. Only a handful of different marine creatures occupied this eco-space, but they provide a window into a Jurassic ecosystem immediately before the arrival of the overlying Sutton stone.

Cramped legs and sore elbows inform us soon enough that another option may be exercised in a standing position in search of fossils within the Sutton stone. Just here in zone A (fig. 7.1b), there are few body fossils among the pebbles and cobbles that compose the conglomerate, among them oysters. Our time is well invested in a closer examination of the cobbles eroded from the parent Carboniferous limestone. One by one, all the same biological elements identified as living on or within the rock bed are found encrusted on individual cobbles or bored into the same cobbles. In whatever way the Sutton stone arrived in this place to bury swales in the underlying Carboniferous limestone, the same diversity of marine life took advantage of the constituent cobbles to maintain a toehold on the environment. To be sure, once the Sutton stone began to accumulate, the sea life fixed to the surface of the Carboniferous limestone was extirpated. Likewise, life clinging to individual cobbles within the Sutton stone was extinguished with the arrival of more and more conglomerate added from above. The crucial question is: How quickly did the Sutton stone accumulate?

More exposures of the contact between Carboniferous limestone and overlying Jurassic limestone await along the coast to the southeast in zone B (fig. 7.1). The change in location from one zone to the next brings us downward through parts of the Sutton stone not viewed earlier and ever closer to the present-day beach, a difference in elevation of more than 23 feet (~7 m). In effect, we explore the older part of a steep island shore when the Jurassic sea level was at a lower datum. Initially, the Sutton stone appears much the same as before. But here, it is possible to find larger coral colonies (*Phacelostylophyllum* sp.) preserved in life position and spread out over a lateral distance of 8 feet (2.5 m). These are recognized as phaceloid in morphology,[9] meaning that individual polyps dwelled in calcified tube-like structures growing subparallel to one another but reinforced as a single colony by calcified struts or crossbars. Compared to their overall bulk, such corals are fragile due to the amount of open space between each tube. On average, the colonies are 12 inches (30 cm) in diameter and attain a maximum height of 8 inches (20 cm). They are 50 percent larger than the corals observed earlier on the former seabed in zone A.

Beneath, there appears a layer of limestone composed solely of sand-size carbonate grains with some crossbedding, and below that, a layer of coquina consisting more-or-less entirely of disarticulated clam shells.[10] The base of the cliff that is still accessible to us during low tide consists of breccia dominated by crudely broken blocks of eroded Carboniferous limestone that are closely packed with little intervening sediment. Some are quite large, as much as 6.5 feet (2 m) across. Must the full sequence of strata combined between the two zones at different elevations (zones A and B) be regarded as part and parcel of the Sutton stone?

It is a question we may ponder during our lunch break, seated higher in the section as the tide continues to rise in the Bristol Channel.

Technically, the name for the Sutton stone represents the entire succession of different layers we've been able to review, but the environment went through subtle changes in water depth and intensity of energy as those layers accumulated one after the other. The blocky conglomerate at the bottom of the pile is analogous to the quartzite coast in Churchill, Canada (chapter 3), where great boulders were eroded from the parent body at the end of the Ordovician period. Similarly, the megabreccia at the bottom of the Jurassic stack represents severe erosion of the Carboniferous limestone parent, most likely under intertidal conditions. The succeeding coquina was formed by clam shells that accumulated in shallow, subtidal water due to currents that swept together empty shells. The following layer of carbonate sand likely formed under less energy but in water shallow enough for currents to move small-scale submarine dunes. The mixed conglomerate and carbonate sand at the top of zone B, also found across the top of the Carboniferous limestone at zone A (fig. 7.1b), require more careful thought.

Meanwhile, a more pressing question must be confronted. The Jurassic period occurred in the middle of the Mesozoic era and lasted 54 million years. Whether or not we consider the full succession of Jurassic strata at Pant y Slade or only the more classic conglomerate of the Sutton stone, we need to know what part of the Jurassic Period we've found ourselves in. Given that Triassic rocks are present between the Carboniferous and Jurassic rocks closer to Ogmore-by-Sea, but entirely excluded between the Jurassic and Carboniferous at Pant y Slade, the nature of the time gap needs to be addressed. The gap could be even greater than the missing Triassic, depending on what part of the Jurassic System (lower, middle, or upper) is represented by the Sutton stone. If much of the Triassic System is locally cut out, so could parts of the Jurassic. It is something to be considered on the path to Dunraven Bay (fig. 7.1b).

None of the Jurassic fossils encountered so far are especially useful in delimiting Jurassic time. A good index fossil is needed, and within the Jurassic system, ammonites are the best evolutionary clocks. The name Ammon comes from the ram-headed Egyptian god of pharaonic times, and ammonites are the coiled shells that belonged to squid-like animals with a shape superficially like a ram's horns. Many Jurassic ammonites are small, a little larger than a quarter, but others are quite a bit larger. Before the group went extinct near the end of the Cretaceous period, the largest ammonites were the size of an automobile's wheel. Large or small, ammonites are seldom plentiful at any given interval of Mesozoic strata. We could spend a week hunting for ammonites in the Sutton stone, or other layers close above. In his classic study of the region, Trueman described an ammonite now assigned to the species *Schlotheimia angulate* in the last of three ammonite zones that delimit the Hettangian stage of the Jurassic system.[11] Referring back to the geological time chart in chapter 1 (fig. 1.9) the Hettangian stage appears at the base of the Jurassic system,[12] one of eleven such stages with an average duration of roughly five million years. The oldest part of the Jurassic system is marked by ammonite zones not recognized at Slade's Hollow on the shores of the Bristol Channel. However, the Welsh sequence that includes our ecological snapshot of Jurassic life on a Carboniferous rock bed is close enough to the start of the Jurassic period 199 million years ago.

The path eastward from Pant y Slade brings us across a fault (fig. 7.1b), where the Jurassic Porthkerry Formation is dropped down in position relative to the older Sutton stone. Here, we find interbedded shale and limestone character-istic of Britain's classic Blue Lias. Fossils especially common in the shale beds include a boat-shaped bivalve (*Gryphaea* sp.) along with sea urchins and rare ammonites. Branching, horizontally oriented burrows (*Thalassinoides*) made by an animal much like today's ghost shrimp are prevalent at some horizons. The setting reflects a fully off-shore environment well away from the rocky coast of a Jurassic island. Beyond the car park overlooking Dunraven Bay, we may continue farther to the headland at Trwyn y Witch. To the southeast, the unbroken view across towering rocky shores provides a sense of the massive Blue Lias. The 2.2-mile (~3.5-km) hike from the Sutton car park has taken us through a remarkable journey, one that bridged a time gap between Triassic and Jurassic rocks and an even more substantial gap between Carboniferous and Jurassic rocks.

THE STORM OVER JURASSIC STORMS

The tenure of two professors from the nearby University College of Swansea brought significant advances to how the geology at Ogmore-by-Sea has been interpreted. Arthur E. Trueman taught at Swansea from 1920 to 1933, which gave him ample opportunity to study the local geology. Trueman mapped a cluster of Jurassic islands between Bridgend and Dunraven Bay and hunted the ammonites from marine strata around those islands to first show their relation-ship to the Hettangian stage.[13] The paleoecologist Derek Ager took a teaching post at Swansea in 1969 and bestowed the name St. David's Archipelago on the Jurassic islands, named after the patron saint of Wales. Controversy arose when Ager (1986) offered a startling interpretation of the Sutton stone as a major hurricane deposit. In perhaps the shortest conclusion to a scientific paper ever written, he proclaimed: "It all happened one Tuesday afternoon." Opposition to Ager's interpretation appeared in a subsequent publication in the same journal, which triggered a lively debate with echoes that long reverberated.[14]

Various rationalizations can be made for and against the hurricane hypothesis. The first accommodation is that Ager's storm bed refers exclusively to that part of the Sutton stone related to the conglomerate beds and not the full sequence, bottom to top. In particular, strata below zone B are stacked against the side of a limestone sea cliff no less than 23 feet (7m) high. The burial of the cliff followed a natural progression from intertidal breccia to shell coquina to a calcareous sand deposit during a relative rise in sea level.[15] Those changes occurred in succession as the paleoisland foundered. During the last stages of rising sea level, preexisting rock benches became flooded one after the other at the top of zones B and A. The conglomerate beds overlying the rock benches are peculiar for the diffuse concentration of eroded Carboniferous limestone clasts blended with fine lime sand. It was this characteristic that prompted Ager to advance his storm hypothesis.[16]

No less than two pauses in the accumulation of the Sutton stone conglomerate can be demonstrated. In order of occurrence, the first is at zone B where the horizon with phaceloid corals (*Phacelostylophyllum* sp.) preserves whole colonies in a growth position within the conglomerate well above a rock bench. Any disruption by storm waves should have caused the delicate corals to be overturned and broken. Instead, they reflect a quiet ecological moment in time. Likewise, the flooding of the zone A bench denotes an interval of calm during which the rock surface was colonized by a wide range of different marine invertebrates before burial by more conglomerate. In short, the entire bulk of the Sutton stone as separate products of sand and conglomerate could not have arrived all at once.

The upper rock bench with its distinctive swales (fig. 7.2) is likened to the scalloped surface on contemporary rock platforms adjacent to limestone sea cliffs around parts of Puerto Rico in the Greater Antilles.[17] At least in part, those modern-day channels are sculpted by the rhythms of the tides, and marine organisms are free to colonize the surface. In 2017, Puerto Rico was hit by Hurricane Maria, one of the most devastating storms to strike the region.[18] It was a Category 5 hurricane with wind speeds up to 175 mph (280 km/hr). Such an event is capable of suddenly blanketing a bare rock platform with the mixed debris of eroded limestone cobbles and carbonate sand, but no one has bothered to check if this might have happened. Hurricane Donna, which made landfall in the Florida Keys in 1960 as a Category 4 hurricane, with wind speeds up to 145 mph (230 km/h) resulted in physical changes that were closely documented in the aftermath.[19] For example, that storm left a 700-foot (213-m) long and 5-foot (1.5-m) high lobe of lime sand on Sandy Key. The essential criticism of Ager's storm hypothesis comes down to the argument that the Sutton stone was the amalgamated result of several storms, not just one on a single Tuesday afternoon.

REGIONAL PALEOISLANDS WITH THE SAME UNCONFORMITY

The potential severity of storms that affected St. David's Archipelago is best viewed in the context of regional and global Jurassic paleogeography. The Jurassic islands were clustered among several larger islands to the east (Anglo-Brabant Land), southwest (Cornubia), northwest (Welshland), and north (Scottish landmass),[20] all straddling a paleo-latitude of 30° N (fig. 7.4). By comparison, modern-day Puerto Rico sits at a latitude of 18°N and the northern-most of the Florida Keys is at 25.5°N. All islands in the Lesser and Greater Antilles, as well as those of the Florida Keys, reside geographically within the seasonal cone of hurricane activity. Further insight is gained by considering other Jurassic islands in the British Isles.

The nearest of these is found across the Bristol Channel in the Mendip Hills of Somerset. Coastal exposures like that at Ogmore-by-Sea are unavailable for study, but various rock quarries reveal the contact between Carboniferous limestone strata and overlying Jurassic limestone belonging to the Hettangian stage. One such quarry (fig. 7.5a) demonstrates the extreme disruption of the boundary as an angular unconformity. Unlike the junction of flat-bedded Carboniferous and Jurassic strata in zone A at Pant y Slade, Somerset's thick-bedded Carboniferous limestone is tilted at an angle of around 45° or more. During the millions of years between the intersection of those two systems, the original flat-lying Carboniferous limestone was tectonically squeezed to induce tilting before the undisturbed flat-lying Jurassic rocks were deposited on top. Not only were the Carboniferous layers tilted but enough time passed before the ruptured surface was planned flat by erosion before the arrival of the first Hettangian sediments.

However brief, ecological time is expressed at the junction where marine organisms colonized the beveled surface. Superb cross-sections through the Carboniferous limestone (fig. 7.5b) show the deep penetration of boring polychaetes (*Trypanites* isp.). At the same time, a dense cover of oysters (*Ostrea* sp.)

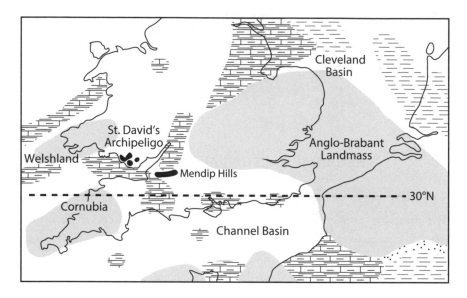

FIGURE 7.4 Map of England and Wales showing the location of the Jurassic St. David's Archipelago among larger paleoislands such as Welshland and Cornubia straddling the paleo-latitude at 30° N.

became attached to the exposed surface (fig. 7.5c). Here in this part of Somerset, there is no equivalent of the Sutton stone conglomerate. It means that once the tilted layers of Carboniferous limestone were first buried, they could not provide the raw material for the erosion of a limestone-based conglomerate.

The Isle of Skye in Scotland's Inner Hebrides provides additional insight into another Jurassic island. In this case, the Hettangian Broadford limestone was deposited on tilted limestone belonging to the Cambro-Ordovician Durness Group.[21] The earliest Jurassic sediments were washed ashore against a hillside formed by the older strata. Fissures in the Cambro-Ordovician limestone are narrow and linear in alignment with the parent strata, penetrating as deep as 16 feet (5 m). Oval-shaped borings occur near the top of the fissures, likely produced by the same boring clams in evidence at Pant-y-Slade (*Gastrochaenolites* isp.). The fissures are filled with eroded fragments of the Durness limestone and the debris of Jurassic shells, including the boat-shaped clam (*Gryphaea arcuata*), sea urchin plates and spines, brachiopod shells, the spicules of calcareous sponges, and phosphatic fish scales.[22] Layers of muddy limestone overlie the unconformity with clusters of *Gryphaea* followed by more than 33 feet (~10 m) of wavy-bedded alternations of silt and muddy limestone. Nothing equivalent to the Sutton stone is evident on the Isle of Sky.

FIGURE 7.5 Details from a quarry section on a paleoisland in the Mendip Hills of Somerset: a) angular unconformity between dipping Carboniferous limestone and overlying, flat-bedded Jurassic limestone; b) Jurassic trace fossils (*Trypanites* isp.) bored by polychaetes in Carboniferous limestone; c) encrustations of Jurassic oyster shells (*Ostrea* sp.) on the unconformity surface of Carboniferous limestone.

All photos by author.

HISTORICAL VIGNETTE: EARLY NINETEENTH-CENTURY PROFESSIONALS AND HOBBYISTS

In his essay on walking, Henry David Thoreau commented about the state of wilderness in England: "There is plenty of genial love of Nature, but not so much of Nature, herself." The statement is a commentary on the intense reordering of the landscape under a long history of human habitation. Every bit of territory throughout the British Isles has come under the exacting scrutiny of naturalists many times over. How could anything be missed? The cultivation of dedicated hobbyists alongside the first true professionals is an important factor. Henry Thomas De la Beche (1796–1855) resides first-in-rank as the English geologist and paleontologist who, in 1835, became the salaried director of the new Geological Survey of Great Britain. Issued as a survey memoir in 1846, a detailed report on the geology of South Wales and adjacent England includes an overview of Glamorganshire.[23] Among the highlights is a sketch (fig. 7.6) of the unconformity at Dunraven Bay, which is safely accessible only during extreme low tides. The scene compares well with localities in the Mendip Hills (fig. 7.5a). Moreover, De la Beche (1835) took pains to illustrate the colonization of Jurassic marine organisms on the Carboniferous substrate, including boring clams and polychaetes (now formally identified as the trace fossils *Gastrochaenolites* and *Trypanites*, respectively). Apropos to fissures in Carboniferous limestone filled by conglomerate eroded from those rocks later on in Jurassic time, De la Beche recognized the comparable phenomenon.[24]

Significant contributions to the paleontology of Glamorganshire were made by hobbyists who combed the area for discoveries in the years following the publication of De la Beche's report. For example, the Bath Natural History Club records show that Charles Moore led a field excursion to Southerndown in 1866 to collect fossils from the Sutton stone.[25] In the years afterward, hobbyists published in the proceedings of the Geological Society of London regarding not only the overall diversity of marine invertebrates in the Sutton stone but also encrusting corals both on eroded limestone cobbles within the Sutton stone and

FIGURE 7.6 Reproduction from the 1846 *Memoirs of the Geological Survey of Great Britain* by Henry Thomas De la Beche, showing the angular unconformity between Carboniferous and Jurassic limestone layers at Dunraven Bay.

Jurassic corals encrusted on the underlying Carboniferous limestone. As such, important clues on Jurassic invertebrate ecology were noted early on by individuals dedicated to the pursuit of natural history.

Another historic locality for Lower Jurassic strata from the Blue Lias occurs in Lyme Regis, a coastal town located some 27 miles (45 km) south of the Mendip Hills on the English Channel (fig. 7.4). No account of British hobbyists turned professional paleontologists is complete without mention of Many Anning (1799–1847), who discovered the first complete skeleton of an ichthyosaur in these Hettangian strata as a twelve-year-old girl. A dolphin-size marine reptile, the ichthyosaur is notable for delivering its young by live birth based on the preservation of fetal skeletons. A contemporary of De la Beche but from a lower social class, Mary Anning went on to achieve a remarkable career as a professional excavator who discovered the complete skeletal remains of the first-known plesiosaur and first-known pterodactyl. Her specimens can be viewed today in the halls of the Natural History Museum in London, where her portrait also hangs in acknowledgment of her skill as a talented fossil hunter. There is little doubt that St. David's Archipelago was bathed in waters where the ichthyosaur and long-necked plesiosaur swam and the great flying reptiles called pterodactyls once nested.

GLOBAL JURASSIC GEOGRAPHY

Missing from nineteenth-century advancements made by the likes of De la Beche, Anning, and countess others of lesser fame, is any notion of how prehistoric coastal life might have been imperiled by sea storms. Francis Beaufort's exemplary effort from direct observation to scale wind and wave conditions stirred by storms at sea had yet to impact the way geologists thought about the accrual of rock formations. The ultimate appraisal regarding the vulnerability of paleoislands like those of St. David's Archipelago to hurricanes comes from global geography circa 199 million years ago. The relationship of Jurassic megacontinents (fig. 7.7) is little changed after the passage of nearly one hundred million years since the dawn of the Permian Artinskian age (compare with fig. 6.8). The Paleo-Tethys Sea remains a prominent feature that straddles the paleoequator, but greater inroads are apparent where the Mediterranean Sea will eventually push through between Africa and southern Europe prior to the initial opening of the North Atlantic Ocean in later Jurassic time. Most notably, the fracture of Pangaea with tectonic activity opening the North Sea through the Cleveland

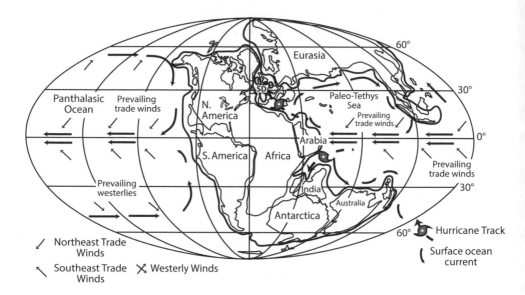

FIGURE 7.7 Global reconstruction for the Early Jurassic (Hettangian) world with interpretations for trade winds convergent on the paleoequator, high-latitude westerly winds, surface ocean circulation, and predicted hurricane paths. SD = St. David's Archipelago.

Basin (fig. 7.4) continued since later Permian time with implications on the development lands of the present British Isles.

Theoretical models for hurricane tracks in the Jurassic Paleo-Tethys Sea predict that major storms should arise seasonally in both the Northern and Southern Hemispheres.[26] Hurricane deposits have not been found in the Jurassic of Arabia or India. However, the predicted pathway for hurricanes in the northern Paleo-Tethys tracks to a geographic cul-de-sac north of 30° latitude in present-day South Wales (fig. 7.7). Although the Paleo-Tethys was much smaller than today's Pacific Ocean, a reasonable comparison also can be made with the cul-de-sac of the Yellow Sea between Korea and mainland China, where direct hits by typhoons are known to impact Hongdo at an even higher latitude at 34°N (see chapter 1, fig. 1.6). Pacific typhoons and Atlantic hurricanes make landfalls more commonly at lower latitudes to strike islands like Taiwan at 24°N or the Florida Keys at 25.5°N latitude. The rare event every one hundred years on the scale of human time amounts to something quite different as an effective agent of change through geologic time over one or two million years. Hence, the landfall of hurricanes in the Jurassic St. David's Archipelago is a justifiable hypothesis.

SUMMARY

Jurassic paleoislands like those of St. David's Archipelago in South Wales, the Mendip Hills of Somerset, and exposures on the Isle of Sky in Scotland's Inner Hebrides encapsulate major time gaps in relation to the older, foundational rocks on which they are built. Sedimentary rocks above and below the unconformities are much the same as variations on the theme of limestone, but millions of years elapsed when the last details of the older and those of the younger limestones were set in solid rock. What happened in those places during times between is fully erased, with no local record of the mass extinctions that changed the course of life. However, the unconformity surfaces yield a treasury of details in ecological time regarding the colonization by Jurassic marine organisms, which included a wide range of new corals, mollusks like the elbow shell, and rock-boring bivalves and marine worms. Captured in sharp focus, such events occurred over brief spans lasting only tens of years at most. The St. David's paleoislands, in particular, also record an amalgamation of storm events mashed together in time but yielding an equally fascinating window into the past. Rachael Carson is perfectly justified in her admonition to reflect on the long history of the blue planet and its susceptibility to big storms that come and go.

CHAPTER 8

HOW ISLAND LIFE ALIGNS WITH GLOBAL CURRENTS

A Journey in Late Cretaceous Time to Baja California's Eréndira Islands

Vast current systems, which flow through the oceans like rivers, lie for the most part offshore and one might suppose their influence in intertidal matters to be slight. Yet the currents have far-reaching effects, for they transport immense volumes of water over long distances.

—Rachel Carson, *The Sea Around Us* (1951)

G eography, climate, and biology contribute to the fundamental aspects of natural history at any place in our world. To the timorous among us, the draw of a peninsula (i.e., almost an island) is irresistible as a solid path thrust outward into the sea. Mexico's Baja California is that slender piece of land, a 685-mile (1,100-km) extension that has attracted me on a regular basis. No part of such an improbably long and narrow parcel of land is far from the sea. On one side, the Gulf of California is normally calm, but on the other, the Pacific Ocean is impetuous with wind and constant sea swells. The coastal borderlands between the United States and Mexico are dramatic for the passage of the California Current, one of those rivers in the sea celebrated by Rachel Carson. The California Current along Alta (upper) California on the U.S. side of the border and Baja (lower) California on the Mexican side is part of a greater oceanic gyre. It originates at a high latitude in the Bight of Alaska, having received the baton in a circuitous race from

Japan's Kuroshio Current. Elements of the California Current track as far south as Ecuador's Galápagos Islands, met there by the north-moving Humboldt Current. The Pacific Equatorial Current is a fusion of the two, continuing westward to complete much of the North and South Pacific gyres driven by trade winds powered by subtropical high-pressure cells centered around 30°N and S latitude (see fig. 4.6).

The modern-day terrestrial and adjacent marine ecosystems of the neighboring Californias are impressive on their own merits. Ashore, great cottonwood trees (*Populus fremontii*) and live oaks (*Quercus agrifolia)* are key members of the native chaparral. Among the other plant species in this regional ecosystem, they are attuned to the passage of wet and dry seasons, also influenced on a multiyear cycle by the arrival of El Niño with its enhanced rainfall. The Californias have an extensive rocky coastline, amounting to 50 percent of overall shore coverage.[1] The dominant member of the rocky-shore ecosystem extending from the Aleutian Islands of Alaska to the southern Baja California Peninsula is the blue mussel (*Mytilus californianus*) that crowds the middle intertidal zone. Secretion of thread-like fibers to secure the shell to a hard surface is critical to the mussel's survival in heavy surf. The nutrient-laden waters of the California Current ensure that the mussel beds are densely populated. Farther out, dense kelp beds of ribbon-like brown algae are the key species in yet another diverse ecosystem. Geologically, these ecosystems are anchored to granodiorite from former magma chambers, now unroofed between Canadian British Columbian in the north to the tip of the Baja California Peninsula. The original magma chambers lay at the roots of Jurassic and Cretaceous volcanoes that spewed lava dominated by andesite, the most common volcanic rock in the Andes Mountains of South America.

Beneath the vestments of living ecosystems in the coastal Californias, there exist relics of extinct Cretaceous systems that are quite different in their natural history. One has only to scratch the surface, in a geological sense, to experience the vibrancy of those precursors. This chapter takes aim at former rocky shores preserved surprisingly intact some 100 miles (~160 km) southeast of the U.S.-Mexican border near the village of Eréndira. Did the Baja California Peninsula or the California Current exist at that time? How different were the Cretaceous ecosystems compared to those presently in place, and how did those older systems function? Elsewhere around the Cretaceous world, what other paleoislands are well preserved? Overall, what do such paleoislands tell us about the biodiversity of rocky shores and their linkage to marine currents swept along by Cretaceous ocean gyres?

FEET ON THE GROUND

The Cretaceous paleoshores of northern Baja California on the Mexican side of the border are unmatched for their pristine state, left mostly untouched by human encroachment. Population centers on the Baja California Peninsula, like Ensenada in the north and La Paz in the south (fig. 8.1a), are large and sprawling, but between them are smaller towns and villages. Roughly 37 miles (60 km) south of Ensenada, a paved road departs from Mexico Highway 1 to reach the coastal village of Eréndira after another 9 miles (15 km). Taking a turn to the north near the coast, the road is unpaved. Fields that spread out from the village are planted commercially, mostly for cash crops like artichokes. Low hills crowd against the seashore farther north, where the agricultural fields disappear. Beyond Punta San Isidro (fig. 8.1b), the road crosses a bridge over an arroyo that runs to the sea through a narrow canyon. A sandy beach provides ample space for roadside camping. This is El Buey (The Ox), and it makes a good staging point from which to explore the area.

Entrained rollers advance off the open Pacific Ocean from the northwest in endless succession. In the strong afternoon sunlight, the blue ocean swells become transparent as they lift in height just before collapsing in foamy chaos. Through binoculars, the rollers take on a stringy golden-yellow look where the kelp beds thrive. The bull kelp (*Nereocystis leutkeana*) is the prevalent brown alga along this part of the coast. It does not live in the intertidal zone like the sea palm in northern California but anchors in deeper water via a holdfast attached to a stone. The algal stipe is like a rope, extending as much as 100 feet (30 m) to the surface, where it terminates in a hollow bladder or peumatocyst 6–8 inches (5–20 cm) in diameter. The float supports a tangle of fleshy blades that suspend to a depth of 10 feet (3 m) or more. During a storm, some huge brown algae are detached and wash onto the beach, typically stripped of their streamers. Given their size, it is astonishing to learn these plants are marine annuals capable of growing from spores to maturity during a single year.

Tides permitting, it is possible to creep cautiously seaward to explore the lower parts of the modern rocky intertidal zone. Here is the interface between prodigious populations of the blue mussel and masses of emerald-green surfgrass (*Phyllospadix scouleri*) that swirl like streamers with the ceaseless energy of the waves. The most voracious predator of the mussel beds makes its living here. It is the ochre starfish (*Pisaster ochraceus*), which typically sports five arms (but as few as four and as many as seven), having a maximum radius of 5 inches (13 cm). Pale orange with clusters of tiny white spines, the coloration of the starfish poses

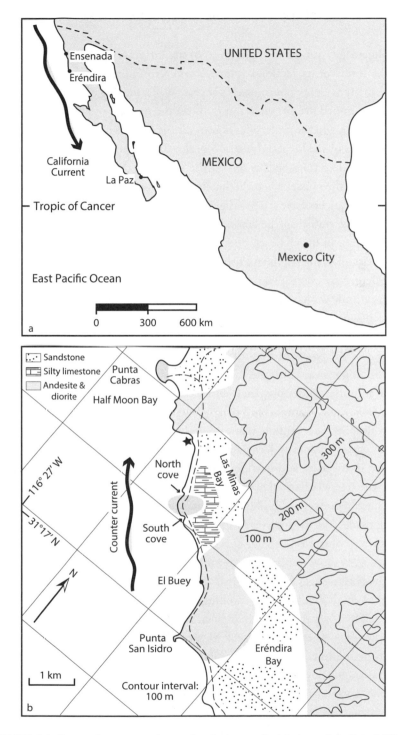

FIGURE 8.1 Geographic, topographic, and geologic maps for Mexico and the Baja California peninsula: a) location map showing the relationship of the Baja California peninsula and California Current to mainland Mexico; b) combination topographic and geologic map featuring the Cretaceous paleoshore and islands north of Eréndira.

a striking image against the bright green of the surfgrass through which it creeps with the surety of its suction-equipped arms. An individual starfish may consume as many as eighty blue mussels per year. When feeding, the starfish assumes a posture humped over its victim. It pulls on the mussel's opposing valves using the suction cups on the undersides of its multiple arms to pry open the shell. The bivalve resists by flexing its adductor muscle to hold its shells locked tight. The first sign of exhaustion spells doom for the victim, as the starfish can extrude its stomach into the mussel when the smallest crack between the two shells opens. The stomach juices of the starfish digest the prey from within.

Never busy, traffic on the coastal road north of Eréndira quiets during the late afternoon. El Buey's beach offers a soft bed on which to roll out a sleeping bag. On cloudless nights, the stars shine with an intensity seldom witnessed elsewhere. Before the evening star rises high above the horizon, one is lulled to sleep by the rhythmic wash of the waves against the shore. At sunrise, a rising tide may leave scant open space on the upper beach, and we repack all our belongings for the day. With a steaming cup of coffee or hot chocolate in hand, there is time to contemplate the power of the surf against volcanoclastic rocks that form tilted layers dipping seaward into the Pacific Ocean (fig. 8.2a). Loading all our camping gear into the van, we prepare to move by stages up the coast to conclude the afternoon at Punta Cabras (fig. 8.1b).

A whopping 140 million years have elapsed since we departed the Early Jurassic coasts of Wales and England, during which fantastic tribes of dinosaurs increased in diversity and quickly dominated terrestrial ecosystems. The plant life on which newly evolved herds of duck-billed dinosaurs and other herbivores fed included conifers, palm-like cycads, and the first flowering plants. Carnivorous dinosaurs like the Jurassic allosaurus and Cretaceous tyrannosaurus dominated the top of the food chain. Our geological clock is now reset to 70 million years ago. Any unsteadiness in the setting by five million years later would bring us to the blue planet's fifth great mass extinction at the end of the Cretaceous period, when all dinosaurs, sea-dwelling reptiles, and flying reptiles of the kind discovered by Mary Anning went extinct. It was a cataclysmic event triggered by the impact of an asteroid that struck Mexico's Yucatán Peninsula and ushered in a deadly period of cold and global pollution by the air-born particulates rich in sulfates that contributed to acid rain.[2]

From the vantage point of a rocky promontory at the side of the road, there is an unobstructed view of the shoreline to the northwest, where a prominent knob of rock rises against the skyline. Less than a mile away (~1.5 km), the hill is our immediate goal for the morning's excursion (fig. 8.2b), where we will set foot on our first Cretaceous island.

FIGURE 8.2 Coastal camping spot near El Buey on the Pacific Ocean shores of northern Baja California, Mexico: a) surf crashing onto Cretaceous volcanoclastic layers; b) view toward the north with wave trains arriving from the northwest (note prominent hill in the central background).

Photos by author.

Isla Las Minas and the South Cove

The hill at Las Minas traces a prominent ridgeline trending perpendicular to the coast and rising roughly 200 feet (~60 m) in elevation. The geological maps at our disposal show that the bedrock exposed along the road at the seaward end of the hill belongs to the Alisitos Formation.[3] Here, it is represented exclusively by volcanic flows of andesite, although Cretaceous limestone is interbedded with those flows at Punta San Isidro. The southeast flank of the ridge is not steep, and the first order of business is to climb to the top, where a 360° view of the surrounding countryside and adjoining seascape awaits. Gazing seaward, the waves below ceaselessly assault the same kind of dark, volcanic rocks on which we stand. Turning to the opposite direction, it is apparent that hayfields flank the ridge on both sides and merge inland around the back side of the prominence. What kinds of rocks lie below the hayfields is not readily discernible, but wave erosion has carved out a pair of coves at sea level to the south (fig. 8.3a) and to the northwest (fig. 8.4a). For convenience, we call them the south and north coves (fig. 8.1b). The geological map informs us that the adjoining Cretaceous strata exposed in the coves belong to the Rosario Formation.

Descending the hill to the hayfield on the east, it is possible to find a few loose fossils scattered in the soil. These are rudist clams, characteristic of Cretaceous strata along a broad swath of the former Tethys Seaway stretching from Oman on the present-day Persian Gulf to Egypt on the Mediterranean to Cuba at the side of the Caribbean. Rudists became extinct near the close of the Cretaceous period (see fig. 1.9). They are like no other mollusks that came before or after in regard to their peculiar subtropical dominion. The fossils from the hayfield are merely fist-size, but Cretaceous rudists I've seen outside Cairo are the size of cannon barrels. A rudist is a mollusk that evolved from an equivalved ancestor to a grotesquely disproportioned clam with one valve shaped like a cone (or barrel) and the opposite valve fitted like a hinged lid. Smaller rudists look something like an ice cream cone with a protective cover, whereas larger rudists call to mind a heavy German beer stein with an ornate lid. The main architectural difference between a beer stein and a rudist is that the stein's lid typically comes with a thumb piece to help the owner open the lid. The rudist animal exhibits no such exterior feature. Instead, it reveals an internal scar near the top of its larger ventral valve where a strong adductor muscle could pull the smaller dorsal valve shut. Another key difference from a beer stein is that the ventral valve is filled almost solidly with calcium carbonate precipitated as the animal matures. Naturally, the ventral valve is much heavier compared to the dorsal valve. Effectively, rudist clams evolved to take over a niche, something like that of coral reefs. The density

FIGURE 8.3 Details from the Cretaceous Las Minas paleoisland: a) view from the top of the island toward the south cove; b) andesite boulder with multiple attachment scars from rudist bivalves in the south cove; c) adult rudist bivalve (*Coralliochama orcutti*) from the south cove (museum identification UABCFCM 0085); d) andesite pebbles encrusted with coralline red algae in the form of rhodoliths at the south cove; e) single rhodolith cut apart to show an andesite pebble fully encased by coralline red algae (*Sporolithon* sp.) from the south cove (museum identification UABCFCM 0077).

All photos by author.

FIGURE 8.4 Additional details from the Cretaceous Las Minas paleoisland: a) view from the top of the island toward the north cove; b) basal conglomerate with eroded andesite boulders at the north cove; c) oyster valve (*Ostrea* sp.) attached in life position to a boulder at the north cove; d) pectinid bivalve (*Lyriochlamys* sp.) found wedged in life position among boulders at the north cove.

All photos by author.

of the larger ventral valve helped the animal to maintain stability against ocean swells, and the largest cannon-shaped rudists grew in close quarters, typically oriented outward in the direction of oncoming waves. Fundamentally, rudists were clams, and some of them fell into a life pattern much like reef corals.

Within the south cove (fig. 8.3a), Rosario strata are found as thick as 30 feet (~9 m), depending on the level of the tides. Here, the formation's basal beds consist of conglomerate with small boulders as much as 16 inches (40 cm) in diameter eroded from the andesite shores of the paleoisland. Superincumbent layers change to silty limestone, tilted southwest off the flank of the partially buried hill. Among the boulders is an exceptionally large piece of andesite that bears attachment scars where the ventral valves of rudists were cemented in place on the rock surface (fig. 8.3b). Fossil rudists from the south cove are the size and shape of an ice cream cone (fig. 8.3c) with a lid some 2 inches (5 cm) across. They were not in the same league as the cannon-size rudists from Egypt. Nonetheless, these Mexican clams contributed to the structure of a modest fringing reef due to their attachment to a rocky shoreline.[4] Projected above a rocky substrate and stabilized by their firm attachment, these rudists formed a secure framework that acted to break the surge of oncoming waves.

Even more abundant than fossil rudist clams, the principal paleontological imprint on the south cove derives from coralline red algae encrusted around every small cobble and pebble of eroded andesite. Layers 6.5 feet (2 m) above the contact with andesite show abundant pebbles in cross-section encircled by a distinctive white rind, often with a pustular surface texture (fig. 8.3d). *Rhodolith* (i.e., red stone) is the term applied to encrusting algae, which are not rooted to the sea floor by a holdfast like the modern kelp but occur as a kind of nodule that freely rolls around on the seabed under the influence of the waves. Here, nodules are commonly 3/4 inch (2 cm) in diameter with a rind completely enclosing an andesite pebble half the diameter (fig. 8.3e). Gradual growth by means of successive thin layers of high-magnesium calcium carbonate secreted organically around a rock nucleus could only be achieved by the constant rotation of the rhodolith subjected to wave activity. Like all algae, these require sunlight to make a living through photosynthesis. Their spherical shape indicates that each part of the rhodolith's outer surface received the same amount of sunlight over the time it grew. Under high magnification, thin sections cut from the south-cove nodules reveal the distinctive reproductive cells diagnostic for a particular genus of coralline red algae (*Sporolithon*) that was abundant during the Cretaceous and remains well represented in shallow seas today.[5] Overall, the fossil rudists and rhodoliths from the south cove point to an intensity of wave agitation that was persistent in real ecologic time as Rosario strata accumulated against the flank of a small paleoisland.

The North Cove

The distance along the road between the south and north coves is barely 550 yards (~500 m), and the van is repositioned for our convenience. Crossing through the paleoisland, it is hard to know how much of the former island's front end has been truncated by present-day erosion. Based on flanking sedimentary deposits around all but one side, the island was larger than found today. The north cove exposes a thick succession of basal conglomerate and silty limestone belonging to the same Rosario Formation, here tilted off the island's flank in the opposite direction to the northwest (fig. 8.4a). Differences on this side of the paleoisland are readily apparent compared to the south side. The most obvious relates to the basal conglomerate, which includes andesite boulders up to 8 feet (2.5 m) in diameter (Fig 8.4b). Small sea stacks attached to the basement rocks also are exposed at low tide.

From a paleontological perspective, the scene is unrecognizable compared to the cove only a short distance away on the opposite side of the paleoisland. The algal rhodoliths so common to the south are nowhere to be found, and rudist clams are scarce. Fossil bivalve mollusks are well represented, but they are dominated by oysters (*Ostrea* sp.) encrusted directly on andesite boulders (fig. 8.4c) and a species of pecten (*Lyriochlamys traskii*) found intact wedged among the boulders (fig. 8.4d). Fossil diversity at the north cove is much greater, with sixteen types recorded.[6] Among them are the delicate moss animals (bryozoans) found encrusted on oysters (*Alderina inuber* and *Cheethamia howei*) or cemented like delicate lace on andesite cobbles (*Frurionella* sp. and *Micropora* sp.). Calcareous worm tubes also occur attached to andesite cobbles. The most common fossils among the larger boulders within the basal conglomerate are regular echinoids (*Trochalosoma* sp.) represented by complete tests. A pencil urchin is recognized based on its robust spines.

Turning inland for a short distance, we cross the road to follow a small gully uphill parallel to the north flank of the paleoisland. Silty limestone enriched by the debris of fossil oysters is exposed within the arroyo. The gully ends 130 feet (~40 m) above mean sea level, but oyster fragments are also strewn through the soil of the adjacent field. At best, the open field extends inland from the coast for a quarter mile (~1 km) before it abuts against andesite bedrock some 330 feet (100 m) above mean sea level. We could repair to the top of the hill for another look, as we did earlier in the morning, but the greater hayfield traces the dimensions of a large embayment within which the Las Minas paleoisland sits bejeweled in its former Cretaceous glory. Las Minas Bay, named for the low-yield iron

PLATE 1 Sea stack called the "Knife" formed by vertically tilted layers of quartzite on Hongdo, South Korea (author for scale).

Photo by B. Gudveig Baarli

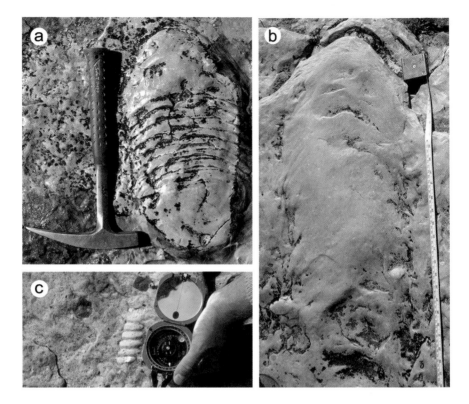

PLATE 2 Megafossils from Upper Ordovician strata exposed on tidal flats during extreme low tide: a) large trilobite *Isotelus rex* (hammer for scale); b) trilobite trackway showing traces left by the forward shuffle of cephalic shield and blunt genal spines. Tape measure for scale extended 40 in (1 m); c) siphuncular fragment of a large orthocone cephalopod (compass case for scale, also recording orientation).

Photos by author.

PLATE 3 Perspective view with detail of the Bater Paleoisland: a) South flank of the complete diorite outlier (dark center) exhumed from overlying Silurian strata to expose the core of Bater paleoisland; b) close-up showing the attachment site with concentric growth pattern of a stromatoporoid colony sitting on the unconformity surface (camera lens cap for scale = 2.5 in or 7.5 cm in diameter).

Photos by author.

PLATE 4 Details of outlying Christopher Bore Paleoisland on the inner (north) flank of the Oscar Range: a) view across the flats that expose Pillara Limestone to the islet's south shore on the horizon (figure for scale); b) view eastward along the islet's south shore with bedded quartzite rising abruptly with physical relief (figures for scale); c) thin limestone layers with stringers of quartz granules exposed adjacent to the paleoshore (pocket knife for scale = 3.5 in or 9 cm).

Photos by author.

PLATE 5 Details from stromatoporoid biozones around the south shore of Christopher Bore Paleoisland: a) oriented *Amhipora* fragments as debris from an inner zone; b) pillar-shaped stromatoporoids in growth position (hammer for scale with 15-in or 38-cm long handle).

Photos by author.

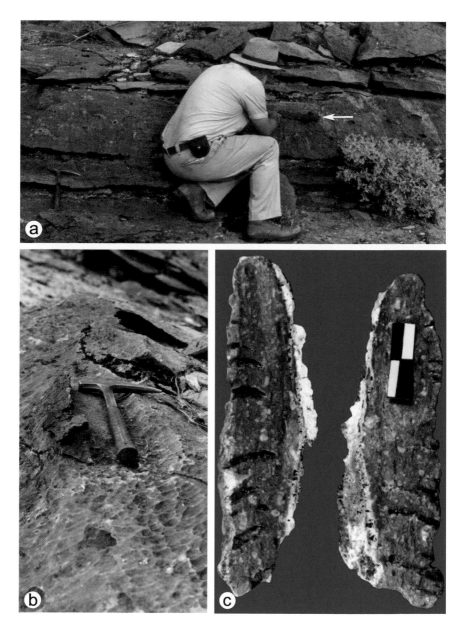

PLATE 6 Details of tower karst near Congo Creek: a) vertical wall of tower karst in Callytharra limestone showing remnant patches of siliceous crust in two places (blue hammer handle in the lower left, marks one of them and white arrow in the middle right points to another); b) close-up view of same siliceous crust marked by white arrow; c) cut and polished layer of limestone from the top of a karst tower, showing subparallel tubes penetrating below the surface (scale = 6/8 inch or 2 cm).

Photos by the author (a, b) and by B. Gudveig Baarli (c).

PLATE 7 Tower karst in Ha Long Bay, northern Vietnam: a) far view showing multiple towers in a maze-like setting; b) closer view of the central tower showing a distinct tidal notch (white arrows).

Photos by author.

PLATE 8 Details from a quarry section on a paleoisland in the Mendip Hills of Somerset: a) angular unconformity between dipping Carboniferous limestone and overlying, flat-bedded Jurassic limestone; b) Jurassic trace fossils (*Trypanites* sp.) bored by polychaetes in Carboniferous limestone; c) encrustations of Jurassic oyster shells (*Ostrea* sp.) on the unconformity surface of Carboniferous limestone.

All photos by author.

PLATE 9 Cretaceous sea stack and related marine deposits at El Volcán south of Eréndira: a) profile of sea stack formed by the andesite Alisitos Formation surrounded by a conglomerate halo in the Rosario Formation (person for scale); b) close view of a small boulder encased in limestone with fossil pecten shell; c) close view of sandy limestone with abundant sea urchin spines farther out from the sea stack.

All photos by author.

PLATE 10 Fossil rhodolith deposit at Cabeço das Laranjas: a) Hill of Oranges locality viewed from an elevation of 280 ft (~85 m) above sea level on Ilheu de Cima; b) profile across south face of the Hill of Oranges showing the burial of a basalt platform by a thick accumulation of rhodoliths.

All photos by author.

PLATE 11 Exposures on the Malbusca coast of Santa Maria Island: a) Pliocene sedimentary sequence sandwiched between pillow basalt with a highly irregular surface (below) and flat-lying pyroclastic deposits (above); b) nearby sea cave at low tide showing extensive encrustation of basalt by coralline red algae.

All photos by author.

PLATE 12 Cleared land showing weathered red clay and soils at Barreiro do Faneca (also known as the Red Desert) exploited by islanders for pottery and roofing tiles (see Locality 7, Topo-map 10.1).

Photo by author.

PLATE 13 Quarry locality north of Ponta das Bicudas: a) limestone shelf 16 ft (~5 m) above sea level, showing protruding knobs of basalt (hammer for scale); b) fossil coral (*Siderastrea radians*) on basalt; c) fossil fire coral (*Millepora alcicornis*) on basalt.

All photos by author.

PLATE 14 Granite coast and fossil coralline red algae on La Dique Island: a) fluted sculpture eroded in granite at Grand'Anse (person for scale).; b) fhinly layered coralline red algae cemented to roof flutes of overlying granite (ruler for scale in inches).

Photos (a) by B. Gudveig Baarli, (b) by author.

PLATE 15 Great Tori Gate at the entrance to the Itsukushima Shrine on Miyajima in Japan's Seto Inland Sea.

Photo by author.

PLATE 16 The great monolith of Uluru (Ayer's Rock) in central Australia: a) aerial view; b) ground view.

Photos by author.

ore once mined in the vicinity, covers an area of nearly 500 acres (~2 km²) and represents one of the region's most clearly defined paleogeographic features.

Crossing Las Minas Bay to Cretaceous Mainland

Returning to the van to continue the day's excursion, it is advantageous to pull off the road a short distance north and look back toward the prominent knoll now on the southern horizon. As we have seen, such a landscape (fig. 8.5a) is not what it seems to be at first sight. The modern rocky coastline on the west side of the road is overprinted by a much older landscape that features its own rocky shores and shallow bays. How old is this Cretaceous landscape, and how might it be correlated with other embayments and islands the same age elsewhere around the world? Based on the disposition of Rosario strata around much of the Las Minas paleoisland, its age is tied to the earliest appearance of those sediments in Las Minas Bay. Indeed, the tilted strata imply that much of the island was slowly buried by Rosario strata during a relative rise in Cretaceous sea level. A general rise in global sea level is thought to have occurred during Campanian time, continuing into early Maastrichtian time near the end of the Cretaceous period (see fig. 1.9).[7] Changes in the magnetic signature of rocks around the transition, also are taken into account regarding global correlation of the Campanian-Maastrichtian boundary.

Global considerations are seldom so easy because the Cretaceous ammonites of Europe are different from those in North America. Correlations within the Upper Cretaceous from the west coast of North America are based on their ammonites and patterns in changing magnetic signatures. Our explorations around Eréndira have not had the good luck to discover ammonites, but the nearest such place is only 12 miles (~20 km) northwest at Punta San José. There, a thick 400-foot (~120-m) coastal exposure of Rosario siltstone includes intervals with abundant ammonites (including the long-ranging *Baculites inornatus*) also found farther north in Washington state. Among other distinctive ammonites, this species is correlated with a particular magnetic anomaly (C33n) that places the sequence within the upper part of the Campanian stage, perhaps some four million years before its end.[8] Unsuited to ammonites, Las Minas Bay was a shallow body of water that long persisted as the bay filled up and Cretaceous time encroached on the Maastrichtian roughly seventy million years ago. The Cretaceous period drew to a close only five million years later (see fig. 1.9) during a phase of global mass extinctions notable for the demise of dinosaurs.

FIGURE 8.5 Aspects of the Cretaceous Las Minas Bay: a) view to the southeast across the raised bay toward Las Minas Paleoisland on the horizon; b) shinbone (tibia) impression from a hadrosaur dinosaur left in place after removal from sandstone strata in the present-day intertidal zone (see fig. 8.1b for field location); c) actual fossil bone now held by university collections in Ensenada (UABCFCM 2612).

All photos by author.

Layers of the Rosario Formation are exposed continuously along this stretch of the road (fig. 8.5a). The thin soil of the hayfield ends at the road, on the opposite side of which there is only bare rock. The unending assault of waves brought by ocean swells from the northwest does its best to undermine the limestone. Looking back across the rocky surface, an occasional plume of spray shoots up into the air like a fountain where seawater caught under pressure by the pounding waves is forced through a small hole in the roof of a subterranean cavity. It is a blow-hole, not an uncommon sight along this shore. It is as if the mighty Pacific Ocean has a faint memory that it once more fully occupied this landscape as it labors to reoccupy the Cretaceous embayment. The volcanic andesite flows that gave rise to Isla Las Minas are more resistant than the limestone. With an ongoing rise in sea level, the rocky knob of the paleoisland may be isolated and resume its former existence as a small island once again.

Back in the van, our route brings us to the far end of the Cretaceous bay, where it abuts a promontory formed by thick andesite flows (fig. 8.2b, see star). The ocean has cut a deep cleft at the junction between sedimentary and volcanic rocks, and we pause here to examine those rocks. The place is called El Destiladero, where a former distillery was serviced by a freshwater spring. The silty limestone more common to the south is replaced by coarse sandstone, some layers of which include pebbles of eroded andesite. It takes a focused search, but the spot where a dinosaur bone was removed from the sandstone left a visible impression (fig. 8.5b). The actual bone (fig. 8.5c) was conserved and sent to the paleontological collections at the Ensenada branch of the Autonomous University of Baja California. It is the shinbone (tibia) from a variety of duckbill dinosaurs, more properly called hadrosaurs.[9] Other reports from this place mention the foot bones of bipedal hadrosaurs. How did dinosaur bones come to be entombed in marine sandstone? Were they washed out of the Cretaceous uplands and into the bay after the animal died? Or did these animals frequent the seashore?

Punta Cabras Paleoisland

The road from Erédira rises steeply after El Distiladero, crossing an andesite ridge to arrive at a side road leading to Punta Cabras (fig. 8.1b). From the descending track, the view over Half Moon Bay and Punta Cabras (Goat Point) captures the stark beauty of a rocky headland linked to the coast by a narrow neck of land (fig. 8.6a). After a lunch pause, the task for the afternoon is to explore the

FIGURE 8.6 Punta Cabras and related physical features: a) general view of the Punta Cabras headland and Half-Moon Bay; b) Cretaceous sandstone from the south side of the peninsular neck of land exposed on Half-Moon Bay; c) deep incision of a narrow channel cut by waves through andesite bedrock on the west front of Punta Cabras.

relationship between the knob-shaped point and its connecting isthmus. The track leads down to a low terrace above the beach on Half Moon Bay. The bay is well sheltered from sea swells striking the headland full-on from the northwest, but lesser waves reach the inner bay by way of coastal refraction. The water is inviting in the midday heat, and a quick swim only sharpens the appetite for lunch.

Upper layers on the neck of land above the beach are formed by Pleistocene dune sand, stabilized by a network of dense root casts. Below is a sandstone layer with large Pleistocene bivalves (*Saxidomus nuttalli*) preserved in life position.[10] Underlying Cretaceous strata are exposed at beach level across the outer length of Half Moon Bay. They consist of coarse-grained sandstone interbedded with andesite conglomerate having a maximum thickness of 8 feet (2.5 m). These beds ramp upward to merge with andesite boulders as large as 6.5 feet (2.0 m) in diameter. Direct contact is exposed with andesite basement rocks typical of the Alisitos Formation.

Turning in the opposite direction, we begin a leisurely hike below cliffs of the Rosario sandstone exposed at low tide to reach the end of the bay where the isthmus meets the mainland. The Rosario sandstone halfway across is distinctly laminated, looking like sheets of plywood stacked one atop the next in an unbroken heap (fig. 8.6b). Paying close attention, it is possible to spot an andesite boulder entombed in the sandstone with a trail of coarse shell debris off to one side.[11] We are on a subtidal Cretaceous seabed midway between a small island and the mainland. The daily rhythm of tides and lesser storms have swept sand into the leeward space behind the island. The lone andesite boulder and its shelly shadow offer a picture in miniature of the greater paleoisland. In this case, it represents a microenvironment wherein broken shells came to rest on the sheltered side of an object too large to disrupt under the flow of water. Standing here on a pleasant spring afternoon in the shade of a sea cliff, a different sensation is associated with a former Cretaceous environment. Here, one can feel the tug of a current streaming over the seabed toward the main coastline yet ahead.

Reaching the shore at the inner termination of Half Moon Bay, the Rosario Formation mirrors the scene witnessed some 330 yards (~300 m) on the bay's outer end at Punta Cabras. Interbedded sandstone and andesite cobbles are roughly the same thickness as before but ramped upward in the opposite direction to merge with andesite boulders eroded from solid bedrock. Fossils among the boulders are reminiscent of the north cove at Las Minas paleoisoand. Cretaceous oysters are attached to cobbles in life position. Remains of colonial corals encrust larger boulders. Shells from the same pecten (*Lyriochlamys traskii*) occur among the rocks. The physical layout of the place is plain to see. What remains to be pondered is the erosive dynamism of waves and currents that segregated paleoislands like Punta Cabras from parent rocks on the mainland. It is not quite the same story told by the quartzite skerries and islets in the Devonian lagoons of Western Australia (see chapter 5). Here, something else has been at work besides waves battering through the walls of folded met-amorphic rocks.

To finish the challenge, we return to the terrace where the van is parked above the beach and continue on foot across the knob of Punta Cabras to confront the open Pacific Ocean. The headland covers an area of 60 acres (~0.25 km²), and the Alisitos Formation at the tip of Punta Cabras is formed entirely of andesite with a pattern of intersecting joints. Natural jointing in volcanic rocks develops after a surface flow stops and begins to cool. The cooling may take place inland long before those rocks might be subject to erosion by ocean waves. Intersect-ing joints are evoked by the shape of sea stacks separated from the front of the Cabras headland, where the solid rock has been excavated from lanes where the joints meet.[12] Essentially, Punta Cabras is its own minor mainland, from which sea stacks have been and continue to be chiseled away.

In this process, ocean waves collide violently against the naked andesite forc-ing water under pressure into the smallest partings in the joints. Where joints are closely spaced parallel to one another, wave action functions like a hydrau-lic jackhammer that wedges into openings to loosen those parts with the most densely spaced joints. Once a piece of rock is broken free and carried away by the waves, its rough edges are subject to rounding by constant collision with other such pieces until smooth cobbles emerge that may be buried beneath other cob-bles in the foreshore. The crossing of vertical joints is the most effective pattern that leads to the isolation of sea stacks, but sometimes a single set of uniformly oriented joints is dominant. The most astonishing natural feature at Punta Cabras is Keyhole Beach, named after its map outline tracing the shape of an old-fashioned key with a narrow slot below a larger, more circular opening for the key stem. The slotted inlet is nearly 200 feet (~60 m) long and no more than 30 feet (10 m) wide (fig. 8.6c). The back of the inlet is enlarged and semicircular, where sand with a bluish tint has accumulated from the crushing of millions of mussel shells detached by strong waves from their moorings at the front of the headland. Wave after wave is funneled into Keyhole Beach with much intensity, and they surge through the passage with a roar of white-foamed water. Given sufficient time, the onslaught will cleave Punta Cabras in two parts, separated by a through-going passage.

Imagine a rocky coastline formed by andesite, much like Punta Cabras, stretched clear across the front of Las Minas Bay. By a combination of intersecting joints and other spots with closely spaced parallel joints, much of the embayment around Las Minas paleoisland was gradually worn away by Cretaceous waves. The paleoisland persisted, in large part, because its seaward-oriented face is more robust, with widely spaced joints less vulnerable to assault.[13] As the day fades, a return to the beach at El Buey beckons for another night of camping.

OTHER PARTS OF THE ERÉDIRA PALEOSHORE

The path inland from El Buey leads through an andesite canyon, where it intersects a track running parallel to the coast. Westward, that track rises steeply to a neck of open land at 260 feet (80 m) above sea level. There, it is possible to look back over the low Cretaceous terrain of Eréndira Bay in one direction and forward across Las Minas Bay in the other. Local topography seen from the perspective of the Cretaceous paleoshore, with its abutting sandstone and andesite, underscores what was once a prominent peninsula, trending southward to Punta San Isidro (fig. 8.1b). Following any of the canyons in the backcountry along an elevation of 330 feet (100 m) brings the visitor to a transition with granodiorite that cooled within a buried magma chamber. The experience of hiking the backcountry lends a deeper appreciation for what was an intermittently violent landscape, with active volcanoes that spewed ash and tephra just before incursions of the rising sea in Late Cretaceous (Campanian) time.

Returning by the coastal road to Eréndira village, a sidetrack off the paved road crosses the Eréndira estuary, giving access to the coastal plain on the south bank. The track follows across a Pleistocene marine terrace for 3 miles (~5 km) to a landmark called El Volcán. Indeed, the cone-shaped feature is mapped as a small Pleistocene volcano.[14] On the seaward side of the track below the edifice, a prominent sea stack composed of Alisitos andesite rises 10 feet (3 m) above tide pools. The profile of this Cretaceous feature is handsomely exposed to the core on its southern flank, revealing a cross-section through enclosing Rosario strata (fig. 8.7a). Andesite boulders packed together around the base of the stack give way to layers of coarse sandstone, including diffuse cobbles. The sandy matrix filling spaces around boulders near the base of the stack include the disarticulated shells of pectens (fig. 8.7b) and oysters with valves still united. Sandy limestone at the top of the sequence fails to completely bury the sea stack, leaving its top exposed. Bedding surfaces from the uppermost layers are strewn with the spines of regular sea urchins (fig. 8.7c). The spines reveal a preferred orientation trending N 45°W and S 45°E, essentially parallel to the Cretaceous coastline.[15] Here, one can feel the pull of waves and the influence of a long-ago current that passed this way.

FIGURE 8.7 Cretaceous sea stack and related marine deposits at El Volcán south of Eréndira: a) profile of sea stack formed by the andesite Alisitos Formation surrounded by a conglomerate halo in the Rosario Formation (person for scale); b) close view of a small boulder encased in limestone with fossil pecten shell; c) close view of sandy limestone with abundant sea urchin spines farther out from the sea stack.

HISTORICAL VIGNETTE: CHARLES DARWIN, MARINE IGUANAS, AND SWIMMING DINOSAURS

Among the islands visited by the HMS *Beagle* during its circumnavigation of the globe between 1831 and 1836, Charles Darwin spent more time in the Galapagos Islands (thirty-five days) and devoted more words to its description (11,373) than any of the several other islands treated in the first edition of his famous travelogue issued in 1839.[16] The blue planet's only living species of marine lizard (*Amblyrhyncus cristatus*) was among the fantastic animals observed by the young Englishman during his formative time in those equatorial islands. Large populations of the lizard live along the rocky shores of Fernandina Island (fig. 8.8), known to Darwin by its English name Albermarle Island. The largest individual Darwin encountered was 4 feet (1.2 m) long from its snout to the tip of its tail and weighed 20 pounds (9 kg). In his description of the reptile's habits, he dispelled the notion that the animal was a fish eater, with the results of stomach dissections proving an exclusive diet of marine algae. Moreover, Darwin observed the reptile "swim with perfect ease and quickness, by a serpentine movement of its body and flattened tail."[17]

Well established as predators during the Jurassic period, ichthyosaurs and plesiosaurs (see chapter 7) were allied marine reptiles that persisted into the Cretaceous period. Mosasaurs represent a third group among marine reptiles but are limited to the Cretaceous period. As a general rule, dinosaurs are regarded as strictly terrestrial in habit. In addition to El Destiladero, where the bones of hadrosaur dinosaurs were recovered from marine strata in the Rosario Formation, a locality near El Rosario another 55 miles (~90 km) southeast is better known for dinosaur fossils from the coeval El Gallo Formation.[18] Teeth and bone fragments typical of meat-eating dinosaurs such as *Gorgosaurus* are reported from terrestrial deposits in that region, but more extensive remains are attributed to a hadrosaur species (*Lambeosaurus laticaudus*). Although identification of the genus requires pending verification of skull bones, the secondary name bears an etymology from the Latin (*latus* and *cauda*) for "broad or wide tail."[19] Notably, the dinosaur's vertically elongated spines attached to tail vertebrae, forming a transversely flattened tail capable of side-to-side movements for aquatic locomotion. In addition to tail spines as long as 22 inches (56 cm) that extend equally above and below tail vertebrae, the pelvic bones from this dinosaur are regarded as weakly articulated compared to other duck-billed dinosaurs more conventionally assigned to a terrestrial habitat.

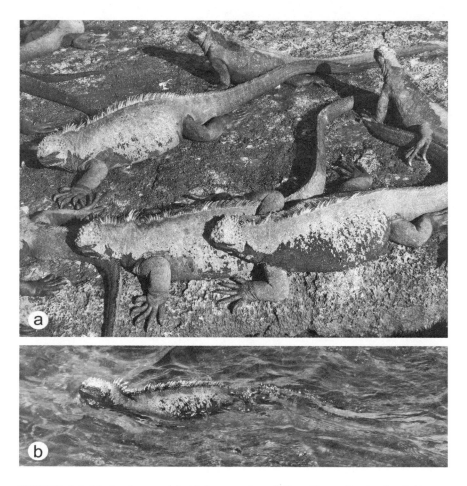

FIGURE 8.8 Marine iguanas (*Amblyrhyncus cristatus*) from Fernandina in the Galápagos Islands: a) cluster of males, about 3.5 ft (1 m) from snout to tail, sunning on the rocks near shore; b) marine iguana swimming.

Photos by author.

The hypothesis of tail-propelled locomotion of water-dwelling dinosaurs has been revived more recently, with studies of Cretaceous fish-eaters belonging to the species *Spinosaurus aegyptiacus* from North Africa with the same kind of elongated tail spines.[20] The plant-eating hadrosaur, which could attain an adult length of 54 feet (16.5 m), was fifteen times the length of an adult marine iguana living today in the Galápagos Islands and would have required far greater resources in terms of aqueous vegetation, whether in the more brackish setting of an estuary or a fully marine habitat. Under that scenario, the population size of such an unusual dinosaur species may have been a limiting factor.

LATE CRETACEOUS GLOBAL GEOGRAPHY
AND BIOGEOGRAPHY

The Tethys Seaway was a distinctive feature at a latitude around 15°N passing between paleocontinents through most of the eighty-million-year-long Cretaceous period. Reinforced by the prevailing westerly flow of equatorial currents in the greater Paleo-Pacific Ocean, the Tethys Seaway accommodated an unusually strong circumglobal current that swept clear around the globe in a westerly direction (fig. 8.9). At a time when the North and South Atlantic oceans were smaller than today, this pattern played a role in the spread of rudist bivalves during the later Cretaceous. It is not surprising to find these unusual fossils well established along the margins of the seaway from one end to the other. What appears odd, however, is that Campanian-Maastrichtian rudists also ranged as far northward as present-day Oregon in North America and as far southward as Peru in South America. Partly based on such biogeographic evidence, a pair of countercurrents is interpreted as divergent from the Tethys Seaway in the western Americas.[21] Because rudist bivalves adopted a sedentary mode of life akin to corals, their chief

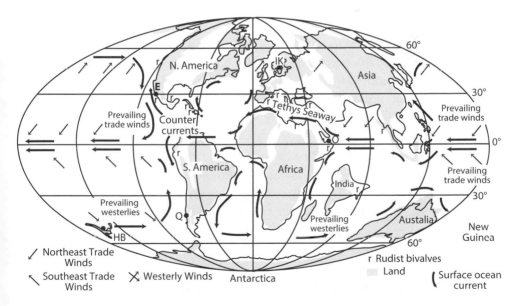

FIGURE 8.9 Global reconstruction for the Late Cretaceous (Campanian/Maastrichtian) world with interpretations for trade winds convergent on the paleoequator, high-latitude westerly winds, and surface ocean circulation. r = rudist bivalves; E = Eréndira (Mexico), Q= Quiriquina Island (Chile), HB = Hawks Bay (New Zealand); IK = Ivö Klack (Sweden) and O = Oman (Persian Gulf).

means of migration from place to place was limited to larval mobility during a free-swimming stage of life. Countercurrents deviating along the shores in the western Americas are a necessary hypothesis to account for the arrival of rudists in Baja California, and farther north to Oregon, as well as present-day Peru. The physical possibility of such a countercurrent compares with the modern Leeuwin Current running from north to south along the shores of Western Australia.[22] That current, which imports warm equatorial water farther south than otherwise normal, is responsible for the foothold of the Ningaloo coral reef that extends for 160 miles (260 km) along the margin of the Exmouth Peninsula. The gyre of the South Indian Ocean is not interrupted by the Leeuwin Current because its north-moving arm with the cooler water it brings from the south is displaced farther off the coast of Western Australia.

The Tethys Seaway likely played a comparable role in the spread of rhodoliths through the Cretaceous world. Studies on fossil and living rhodoliths remain in an early stage of investigation compared to marine invertebrates.[23] Patterns in past biogeography may be inferred through the present-day distributions of the key coralline red algae that produce rhodoliths. In this case, the dominant genus (*Sporolithon*) represented by Cretaceous fossils at Las Minas (figs. 8.3d, e) has surviving relatives in the Mediterranean (one species), the South Atlantic off Brazil (ten species), Eastern Pacific off Mexico (one species), South Pacific in French Polynesia and Australia (four species), and the Western Pacific around southern Japan and Indonesia (five species). The genus is poorly represented in today's Mediterranean region, but links between the highest species diversity in the Brazilian South Atlantic and other Pacific Ocean realms were enhanced by a strong current through the passage between the Cretaceous North and South American paleocontinents. Unlike the rudists, Cretaceous rhodoliths failed to push farther northward than present-day Mexico. The continental seaway that extended north-south from the Caribbean Sea to the Artic Sea in Cretaceous North America (fig. 8.9) lacks any record of rhodoliths, due perhaps to water clouded by muddy sediments. That epicontinental seaway was sufficiently wide to isolate populations of Cretaceous dinosaurs on opposite shores in North America.

OTHER ISLANDS IN THE CRETACEOUS WORLD

Far afield from Mexico's Enéderia, other hotspots of coastal diversity are shown on the Campanian/Maastrichtian world map (fig. 8.9), where fossil remains occur around paleoisoands of greater or lesser size. Two are in the Cretaceous

Pacific basin shared with Eréndira. The margins of Chile's Concepción Bay are ringed by Paleozoic metamorphic rocks and Triassic granite against which the Maastrichtian Quiriquina Formation is emplaced.[24] The name derives from Quiriquina Island at the opening to Concepción Bay, although the focus of paleontological research on the unconformity between metamorphic rocks and overlying conglomerate is nearby on the Chilean mainland. Higher above the unconformity, coarse-grained sandstone and interbedded shell beds that comprise the formation are noted for a diverse fauna that features many mollusks, crustaceans, and vertebrates, including plesiosaurs, turtles, and fish bones. At the unconformity surface with hard metamorphic rocks, the primary evidence for Cretaceous colonization is from rock-boring bivalves (*Gastrochaenolites* isp.). Fossil wood fragments from the basal part of the overlying sandstone also exhibit borings made by another bivalve (also *Gastrochaenolites* isp.).

The densely forested Hawkes Bay region on the North Island of New Zealand includes limited exposures of an intra-Cretaceous unconformity between sandstone in the Urewera Group and an overlying conglomerate belonging to the Maungataniwha Formation. The basal 3.5 feet (1 m) of this unit is reported to yield a rich Campanian fauna, including twenty-four kinds of marine invertebrates and fish vertebrae.[25] Bivalves are especially diverse, with as many as a dozen species, among which tethered types (*Arca* and *Modiolus*) are noteworthy as clams attached to hard surfaces by their tough byssal threads. Barnacles attached to sandstone pebbles also are common.

The Campanian boulder shore at Ivö Klack in southern Sweden is one of the most thoroughly studied fossil rocky shores anywhere in the world.[26] The locality is confined to a small lake island where clay deposits have been extensively mined from a basin lined by large boulders of Precambrian gneiss. Dense encrustations of bivalves, marine worms, and bryozoans are preserved in place on the surfaces of the boulders and show distinct patterns of vertical zonation. Effectively, it is possible to witness the banded zones of marine invertebrates as clearly as if canvassing a modern rocky shore in the water with swim goggles. An upper zone spanning some 16 inches (40 cm) is dominated by large rock oysters (*Spondylus labiagtus*); whereas a partially overlapping middle zone is prominent for larger oysters (*Amphidonte haliotoideum*). The lowest zone, with a spread of some 20 inches (50 cm), exhibits concentrations of an encrusting inarticulate brachiopod (*Ancistrocrania stobaei*) and the calcareous tubes of serpulid worms. Vertebrate remains from associated sediments include the bones of plesiosaurs, mosasaurs, crocodiles, fish, and the teeth of sharks and rays.

The greatest diversity of species from a Maastrichtian shoreline comes from Oman on entrance to the Persian Gulf, where the Late Cretaceous latitude was

near the paleoequator (fig. 8.9).[27] Different settings within the Qahlah Formation entailed colonization of silica cobbles derived from the Semail ophiolite
(uplifted deep-sea deposits) and limestone hardgrounds. The most common is
an oyster (*Acutostrea* sp.), but several kinds of encrusting corals are recognized
in addition to encrusting bryozoans. Limestone hardgrounds were extensively
colonized by rock-boring bivalves (*Gastrochaenolites* isp.), and fossil wood shows
evidence of activity by wood-boring bivalves comparable to the Chilean locality
in Concepción Bay. As in southern Sweden, the tubes of serpulid worms are present, as is an encrusting inarticulate brachiopod (*Discinisca* sp.). Rudist bivalves
contribute to the fauna, also listed as an encruster.

On the whole, an equatorial invertebrate fauna should be more diverse than
one at latitudes closer to 50°N or S latitude in the case of the Ivö Klack and
Hawks Bay faunas. Notably, no fossil corals belong to either of those faunas.
By comparison, the Eréndera rocky-shore fauna is intermediate in diversity but
more dominated by rudist bivalves and the algal rhodoliths.

SUMMARY

Compared to other Cretaceous paleoshores and paleoislands with limited
exposure, the Eréndira landscape of northern Baja California unveils a nuanced
tapestry extraordinary for its fidelity to a roughly seventy-million-year-old
seascape. Through the grind of Late Cretaceous time, what began as a linear
coastline formed by volcanic flows was gradually transformed by erosion into a
more curvaceous shore with indented bays and small islands that fronted stout
headlands. The spacing of intersecting joints in the otherwise resistant andesite determined where the land held firm and where it failed against a rising sea.
Long preceding the arrival of blue mussels adhering to the same rocks, bizarrely
modified rudist clams populated a coast where the stimulation of waves best
suited their colonization. Likewise, the spherical-shaped rhodoliths secreted by
coralline red algae sought the same vigorous tumult of waves long before modern
kelp beds found their anchorage. In more sheltered settings, Cretaceous oysters,
bryozoans, sea urchins, and some corals took refuge in a more sheltered habitat.
The entire Eréndira seascape was brushed by a countercurrent flowing northward
in opposition to the Pacific Ocean gyre farther offshore. The countercurrent
carried the germ cells of marine invertebrates on a side branch of the powerful
river-like highway spun off the Tethys Seaway. Rachel Carson understood the
vigor of the world's ocean gyres, those long rivers flowing at sea, and correctly
surmised their presence in worlds long vanished.

Mexico's Gulf of California would not appear in any configuration like its present form for another forty-five million years, and the mainland supported an entirely different set of animals, including dinosaurs. Preferring the comfort of the soft sand beach at El Buey, I have never gone to the trouble of camping on the paleoisland at Las Minas. In my dreams, however, I sometimes find myself setting up camp on the islet at the end of a Late Cretaceous day. Instead of scanning for the spouts of whales cruising the California Current, I listen intently for the splash of a duck-billed dinosaur in the embayment below. Could they swim? At least they lived nearby, and surely, their vocalizations will punctuate my preparations for the night.

CHAPTER 9

HOW ISLAND LIFE ADJUSTS TO OPPOSING SHORES ON OCEANIC ISLANDS

A Journey in Middle Miocene Time to the Madeira Archipelago

A beach composed of many cobblestones that grind against each other in the surf is an impossible home for most creatures. But the shore formed of rocky cliffs and ledges, unless the surf be of extraordinary force, is host to a large and abundant fauna and flora.

—Rachel Carson, *The Edge of the Sea* (1955)

A scheduled flight from Lisbon to Funchal on the principal island of Madeira requires an hour and a half. After transfer to a local flight, it takes barely 15 minutes to cross over to the neighboring island of Porto Santo. Portuguese sea captain João Gonçalves Zarco reached a safe anchorage here in 1418 after being blown off course during a violent storm. The island's name, "Holy Harbor," was bestowed by a shaken crew grateful for deliverance from an unforgiving sea. Outmaneuvered in competition with Spain for the conquest of the Canary Islands from its indigenous inhabitants, settlement of unoccupied lands in Madeira, and soon after the Azores, was achieved as a bloodless consolation prize. The fabled School of Navigators established by Prince Dom Henrique (1394–1460) at Cape St. Vincent in southwestern Portugal ushered in the age of discoveries, which sent out mariners like Zarco at the start of the fifteenth century CE. In time, Portuguese explorers sailing sturdy caravels capable of tacking against unfavorable winds enriched their small country beyond all expectations through trade routes leading to Brazil,

around Africa's Cape of Good Hope, and onward to India, China, and Japan.[1]
For his part, Zarco remained on Madeira, some 620 miles (1,000 km) off the
Portuguese mainland (fig. 9.1a), where he established the provincial seat of
government in Funchal.

More than seventeen times as large as Porto Santo, the main island of Madeira
has an area of 285 square miles (740 km²), and Pico Ruivo has an elevation
3.5 times as high as Porto Santo (fig. 9.1b). Indeed, Madeira is a volcanic island
of sufficient size and height to regulate its own climate. The name, Madeira,
relates to the thickly wooded terrain dominated by laurel forests with diverse
species among several genera (including *Apollonias*, *Ocotea*, and *Persea*). Fossil
evidence from leaf fragments preserved in volcanic ash shows that an endemic
flora dates back as much as two million years.[2] The existing shield volcano on
the big island evolved from a more explosive stage related to the growth of a
submarine volcano, geological evidence dating to about seven million years ago.[3]
Porto Santo, which exhibits the eroded remnants of a subaerial volcano, transi-
tioned from a submarine state some eighteen million years ago. Madeira and the
lesser island of Porto Santo fall in line with volcanic sea mounts to the northeast
that are yet older, ranging from twenty-two million years nearer Porto Santo to
sixty-seven million years closer to Cape St. Vincent (fig. 9.1a).

Geologic relationships within the Madeira Archipelago and its link to sea
mounts are part of the region's overall story, but this chapter is focused on
the extraordinary rocky-shore deposits on islets associated with Porto Santo.
A combination of marine macrofossils and microfossils found in limestone layers
on Ilhéu de Cima and Ilhéu de Baixo (Portuguese for Upper Isle and Lower
Isle, respectively) delimit a position within the Middle Miocene (Serravalian
stage) dated to fifteen million years ago. The fossil deposits attest to a history of
island submergence and uplift ending in prolonged subaerial exposure with the
build-up of extensive basalt flows and pyroclastic deposits. Based on additional
islets around Porto Santo on all sides like gargantuan sea stacks, it may be estimated that Porto Santo has undergone erosional attrition of about 65 percent.
The northeast trade winds that carried Portuguese sailors to the island have
been in place for a long time, and their attendant sea swells are largely responsible for the gradual diminution of Porto Santo, with its north coast subject to
the greatest impact by rough surf. Like Silurian Bater Island (see chapter 4), the
extent to which windward and leeward differences in wave agitation affected
rocky-shore biotas is of central interest. Rachel Carson spoke to this issue, and
her insights are readily tested by variations in fossil biotas from the Miocene of
Porto Santo. Distinctions between everyday sea swells and major storms also are
part of this story.

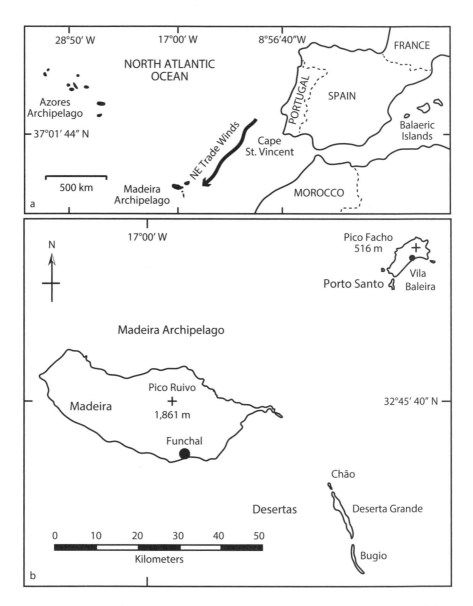

FIGURE 9.1 Maps with a focus on the North Atlantic Ocean and archipelagos settled by the Portuguese: a) Western Europe related to the Madeira and Azores archipelagos; b) islands belonging to the Madeira Archipelago.

FEET ON THE GROUND

The airfield on Porto Santo occupies a wide strip of land through a valley divid-
ing the island into two parts with higher elevations on opposite flanks (fig. 9.2).
Located off to the southwest end of the runway, Pico de Ana Ferreira is among
the island's most visited natural monuments, with a quarry exposure that leaves
no doubt about volcanic origins. Any taxi driver meeting passengers at the airport will know the location and gladly provide a diversionary tour before pro-
ceeding to the settlement at Vila Baleira. The sharply defined and neatly arrayed
columns of basalt represent a former magma pool that filled a large tunnel complex (fig. 9.3). Individual columns form perfect hexagons that are 2 feet (0.6 m)
in diameter and represent the gradual cooling of magma from the sides of the
tunnel to its central axis. Due to an unusually high concentration of oligoclase

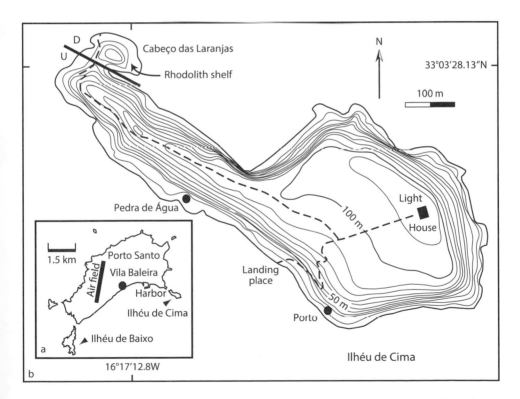

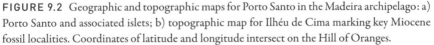

FIGURE 9.2 Geographic and topographic maps for Porto Santo in the Madeira archipelago: a)
Porto Santo and associated islets; b) topographic map for Ilhéu de Cima marking key Miocene
fossil localities. Coordinates of latitude and longitude intersect on the Hill of Oranges.

FIGURE 9.3   Pico de Ana Ferreira on Porto Santo, showing cooling columns formed by basaltic magma with high concentration of oligoclase feldspar (mugearite).

Photo by: B. Gudveig Baarli.

feldspar within the basalt, the peculiar classification of these rocks falls under the name mugearite.

Arrangements are in place for a day-long tour to include three localities with extensive paleoecologic relevance on Ilhéu de Cima, situated less than 2 miles (3 km) from the commercial harbor east of town (fig. 9.2a). The upper isle is a conservation site for which permission to visit is required. Local nature wardens will accompany us on our excursion with transport to and from the island.

Orders are left with a local café for lunch packs to be picked up at 8 A.M., to include Madeira's unique version of pita bread (*bolo no caco*) stuffed with fried beef steak, garnished with lettuce and a fried egg. In the fading light of a midsummer evening, the lighthouse tower sends out a beacon from the island's plateau positioned 407 feet (~125 m) above sea level, easily visible from town.

The morning's plans are met in an orderly fashion. Under a flat sea, the concrete pier near the south end of Ilhéu de Cima is reached by 9 A.M. with a fixed departure time set for 5 P.M. Before climbing the path with switchbacks hewn from the 265-foot (80-m) cliff face that ascends to the plateau, we divert along the shore some 450 feet (137 m) to reach a place identified by name as Porto (fig. 9.2b). Preparing to step onto the paleoshore, our geological clock is advanced fifty million years from the chaotic close of the Cretaceous period (also the end of the Mesozoic era) and well into the following Cenozoic era. The cover of native angiosperm plants on present-day Madeira and that island's fossil evidence are sure signs of major change.

Porto Paleoshore

Vestiges of a Miocene shoreline reveal a fringing coral reef preserved in exquisite detail that abuts basaltic cliffs at Porto.[4] The limestone-covered platform covers roughly 2,000 square feet (180 m²), close enough to sea level that the splash of waves from the last high tide has left puddles (Figure 9.4a). The mind's eye must perform a trick of imagination that requires removing the island's greater bulk rising above, which came into place with sudden and prolonged volcanic violence to extinguish the reef. Densely packed in growth position (fig. 9.4b), thousands of small coral colonies belong to a single species (*Tarbellastraea reussiana*), although each coral head is beveled by recent wave erosion. Nature's designs are captivating, and the similarities between the organic concentration of corals and inorganic repetition of basalt columns at Pico de Ana Ferreira (fig. 9.3) make a strong impression. Many of the Porto corals were infested by boring bivalves, represented by three kinds of trace fossils (*Gastrochaenolites torpedo*, *G. lapildicus*, and *G. vivus*), all of which dwelled within cavities perpendicular to the coral platform. Other corals show borings made by a peculiar pyrogomatid barnacle (*Ceratoconcha costata*) and sponges (*Entobia* isp.).

The reef platform tracks inland for a short space, where it disappears beneath a blanket of water-laid volcanic ash, lapilli (sand-size volcanic ejecta), and basalt (fig. 9.5a). Erosional weathering of the softer ash has exposed complete coral

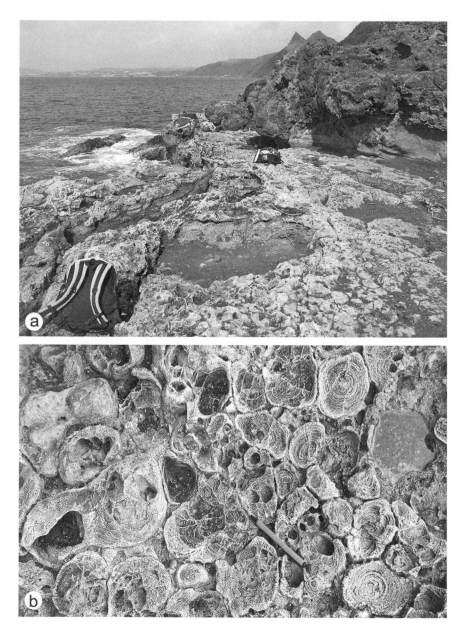

FIGURE 9.4 Miocene limestone shelf at Porto on Ilhéu de Cima: a) general view eastward toward Porto Santo with coral platform in foreground (back-packs for scale); b) close-up showing crowded coral heads in growth position (*Tarbellastraea reussiana*) truncated by erosion (pen for scale = 5.5 in or 14 cm).

Photos by author.

FIGURE 9.5 Burial of Miocene shelf at Porto on Ilhéu de Cima: a) proximal edge of coral platform buried by ash, volcanic lapilli, and basalt (figure for scale with hand pointing to feature shown in following photo); b) well-preserved coral colony in life position (*Isophyllastrea orbignyana*) partially exhumed from beneath a cover of volcanoclastic sediments (pen for scale = 14 cm).

Photos by author

colonies that include another species (*Isophyllastrea orbignyana*), more common during the final stage of reef development (fig. 9.5b). Indentations in the island's vertical cliffs allow for further exploration of the paleoshore, where a low rocky coast rose above the coral platform in places (fig. 9.6a), also buried by the same blanket of volcanic ejecta. It is impossible to discern from a single plane parallel to the coastline whether the Miocene sea cliffs may have reached any higher

FIGURE 9.6 Details of Miocene rocky shore at Porto on Ilhéu de Cima: a) figure stands with back to basaltic sea cliffs showing topographic relief from waist level to head and hand pointing to feature shown in following photo; b) encrustation of coralline red algae with distinctive pustular texture on basalt rock face (pen for scale).

Photos by B. Gudveig Baarli (a) and author (b).

within the core of Ilhéu de Cima. Be that as it may, sufficient exposure is available to distinguish a tripart biological zonation through a vertical height of 5 feet (1.5 m).[5] At the top, basalt pinnacles are covered by crustose coralline algae that retain a distinctive pustular texture (fig. 9.6b). About 5 percent of the exposed former rocky shore exhibits preservation confirming the presence of the red algae at this level, although crusts that are poorly preserved cover another 35 percent of the lower surface. In addition to coralline algae, large barnacles (*Balanus* sp.) and calcareous tubes of serpulid worms are limited to the upper zone. Bivalve mollusks represented by two genera (*Spondylus* sp. and *Lima* sp.) occur cemented in place in a middle zone together with a few corals (*Isophyllastrea orbignyana*). Starting in the middle zone but extending into the lowest of the three zones, bivalve borings preserved as trace fossils (*Gastrochaenolites torpedo*) penetrate basalt by as much as 1.5 inches (4 cm). Abundant, smaller barnacles also encrust basalt in the lowest zone. Such a detailed pattern of biological zonation is seldom encountered in the fossil record, and the Porto example is similar to the older Cretaceous example from Ivö Klack in southern Sweden (see chapter 8). In this case, however, Rachel Carson's prediction holds true regarding the capacity of rocky ledges and vertical cliffs to support a diverse marine biota in the absence of gravel churned by the rough surf.

Cabeço das Laranjas Paleoshore

Stopping at the lighthouse on the island's plateau after a heart-pounding ascent (fig. 9.2b), the opportunity to catch one's breath and enjoy the vista over Porto Santo's south shore brings a welcome respite. Beyond the pier at Vila Baleira, the golden sands of the island's celebrated beach stretch out uninterrupted for 2.5 miles (4 km). It's the chief attraction that draws tourists to the island during the busy summer season. The coastline between Ilhéu de Cima and Ilhéu de Baixo, with the long sandy beach on the mother island, belies the fact that the south shore enjoys a sheltered position leeward from the steady trade winds.

The path from the lighthouse leading northwest across the narrow part of the island brings us to the second locality for the day's excursion: Cabaço das Laranjas (Portuguese for Hill of the Oranges). Halfway along the trail, before a gradual descent from 300 feet (~90 m), a spectacular viewing point overlooks the faulted north end. Our goal is a limestone platform covering 4,845 square feet (450 m²) exhumed from beneath a great knob of columnar basalt (fig. 9.7a). From a distance, the limestone shelf is reminiscent of the Miocene coral platform at Porto, but the exposure is 2.5 times as large, and the content is very different from the corals preserved there in growth position. What awaits is a deposit of rhodoliths similar to those from the flank of the Cretaceous Las Minas paleoisland (see chapter 8), although in this case, the Miocene rhodoliths are the

FIGURE 9.7 Fossil rhodolith deposit at Cabeço das Laranjas: a) Hill of Oranges locality viewed from an elevation of 280 ft (~85 m) above sea level on Ilheu de Cima; b) profile across south face of the Hill of Oranges showing the burial of a basalt platform by a thick accumulation of rhodoliths.

size of large oranges, and they are stained that color due to the leaching of iron compounds. Hence, the colorful name ascribed by locals is most appropriate. The other aspect deserving attention is the level of wave activity on the island's exposed coast under the influence of the northeast trade winds. The reason for our circuitous hike is because a safe landing spot at the north end can't be assured under conditions with a choppy sea.

Rounding the northeast corner of the island, the first fossil rhodoliths are encountered along the trail in a layer at an elevation of 70 feet (21 m) above sea level. The same horizon is closer to sea level on the opposite side of a distinct fault line (fig. 9.2b), indicating a drop of about 62 feet (19 m) from the trail-side strata. Leaving the trail to reach the east face of the fault block, it is clear from its higher elevation (fig. 9.7a) that the Cabaço das Laranjas sequence formed slightly later than the deposits at Porto. The age difference is slight, however, and it comes down to an interpretation that environmental conditions were different for different organisms on opposite sides of the island.

As the noon hour nears, it is tempting to open the lunch packs wrapped to insulate the warmth from the morning's fresh *bolo no caco*. But the limestone shelf at the Hill of Oranges is so extraordinary that our interest delays the lunch break. The actual oranges in our lunch packs are no match in size compared to the orange rhodoliths onsite. Not so clear as viewed from above, multiple rhodolith layers are preserved beneath the basalt hill (fig. 9.7b). The lowest in stratigraphic position is the thickest the most coherent, draped like an apron covering an older basalt formation. Near the fault line, the basal rhodolith layer attains a maximum thickness of 8.5 feet (2.6 m), although it tapers to wedge-like dimensions outward from that point. Succeeding rhodolith layers vary from 5.5 to 24 inches (15 to 60 cm) in thickness, separated from one another by layers of sand and ash. Rhodolith size is variable, but a diameter of 4 inches (10 cm) is common. Analysis of diagnostic characteristics found by Johnson et al. shows that species belonging to at least three genera are present (*Sporolithon* sp., *Lithothamnion* sp., *Neogoniolithon* sp.).[6] The overarching concept to remember is that rhodoliths are precipitated by coralline red algae with growth entirely dependent on sunlight.

In life, any rhodolith trapped at the bottom of a substantial deposit will be deprived of sunlight and perish. Thus, a thick cover of rhodoliths signifies something other than the habitat normally conducive to good health. Unlike the coral platform at Porto, which grew in place, the rhodolith shelves at the Hill of Oranges denote deposits that suffered transport from elsewhere on the sea floor where they once thrived. Annual banding is recorded in the branches that radiate outward from the center of living rhodoliths, and it is calculated that a rhodolith with a diameter of 6 inches (15 cm) represents a life span of 120 years.[7] Few fossil rhodoliths at Orange Hill are so large, but it is unlikely for a rhodolith's algal surface to remain alive beneath a thick deposit of other rhodoliths for decades without exposure to sunlight. If they did not dwell here, how did the Miocene rhodoliths of Orange Hill arrive at this place?

Hunger wins out, and the basalt knobs surrounded by rhodoliths are taken as lunch seats. Different functions may be undertaken at the same time: One can puzzle out questions while also enjoying a good lunch. The knobs that protrude above packed rhodoliths are close together (fig. 9.8a). Some are merely large boulders but based on the profile of the basalt knobs exposed at the platform's edge, others are sea stacks. The surface density of rhodoliths is high, on average 45 per square yard (~1 m²), and the intriguing issue is the depth of a deposit that may easily exceed 5 feet (1.5 m). Essentially, the spacing of sea stacks (not to mention the introduction of eroded boulders) acted as a baffle to entrap rhodoliths once they were swept shoreward from deeper water.[8] Depending on water clarity, living rhodoliths are known to dwell at water depths between 6.5 and 325 feet (~2 and 100 m). The hallmark of rhodoliths is their tendency to assume a spherical shape, and they are easily rolled back and forth on the seabed by the gentle effect of waves and currents. Indeed, if sunlight fails to be absorbed by a given rhodolith over its entire surface during a given lapse of time, it will not take on a spherical shape.

Invertebrates, such as sea urchins, living among the rhodoliths (fig. 9.8b) also are capable of disrupting rhodoliths and turning them over. Many rhodoliths are nucleated around a rock core (fig. 9.8c). However, in deeper water farther away from a rocky coastline, rhodoliths can encrust a bit of shell or broken piece of another rhodolith. It stands to reason that the larger the rhodolith, the longer it must have survived in a favorable habitat. There are fewer fossil rhodoliths with rock cores at the Hill of Oranges than those without. Some rhodoliths exposed at the surface were encrusted by coral colonies after the host stabilized. That is to say, a rhodolith saddled by a coral on one side would undergo a change in its center of gravity and be less able to roll freely.

Ocean swells from the northeast trade winds shift objects on the seabed, but normally at a depth of fewer than 60 feet. (18 m). As swells cross shallower water close to land, they break into violent surf. The energetics of such a setting are tolerated by crustose red algae attached to rock cliffs but are inhospitable to free-living rhodoliths. A hurricane is the likely source of wind energy able to disrupt the seabed, with swells reaching depths inhabited by the largest rhodoliths. The recorded history of hurricanes arriving in Madeira is rare, but one did strike in 1842 and 162 years later with Hurricane Vince in 2005.[9]

FIGURE 9.8 Large fossil rhodoliths at Cabeço das Laranjas (Hill of Oranges) on Ilhéu de Cima: a) pair of Miocene sea stacks partially buried by a deposit of rhodoliths (grid for scale is 19.7 in or 50 cm on each side); b) cluster of large rhodoliths cover an echinoid (*Clypeaster* sp.); and c) single rhodolith in cross section, showing a basalt pebble at its core (pen for scale = 5.5 in or 14 cm).

Photos by author.

Pedra de Água Paleoshore

The day's final stop, at Pedra de Água (fig. 9.2b), is one-fifth of a mile (350 m) along the coast northwest of the landing place. There is no pathway, and the ups and downs over and around obstacles make for slow progress. The site is worthy of comparison as yet a third paleoshore because it combines aspects of the Miocene shores dominated by fossil corals at Porto and by fossil rhodoliths at the Cabaço das Laranjas.[10] Subtle in its paleoecology and easy to overlook, the spot occupies 130 square feet (12 m²) at the base of a high cliff. A cluster of oblong basalt mounds from 6.5 to 26 feet (2 to 8 m) in length having a relief of barely 20 inches (0.5 m) are exhumed from beneath volcanic sediments. The mounds are occupied by coral colonies (*Tarebelastrae reussinana* and *Solenastrea* sp.) preserved in growth position with members of an encrusting rock oyster (*Spondylus* sp.). Etchings by boring sponges left as trace fossils (*Entobia* isp.) are present on some corals, as are abundant borings left by bivalves (*Gastrochanenolites torpedo*).

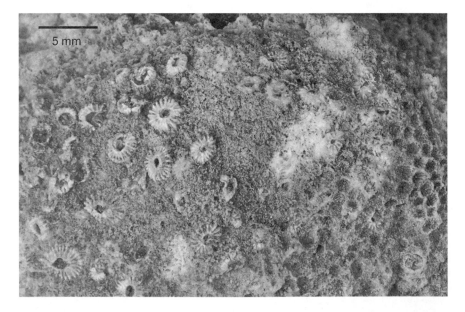

FIGURE 9.9 Surface of coral colony (*Tarbellastrea reussiana*) with multiple examples of a boring barnacle (*Ceratoconcha costata*) visible. Depth of penetration by the barnacles within the coral head is 12 mm, at maximum.

Photos by author.

Exclusive to one of the coral species (*Tarbellastrea reussinana*), the coral-inhabiting pyrogomatid barnacle (*Ceratoconcha costata*) expresses a clear host preference (fig. 9.9). Some oysters are further colonized by crustose coralline algae and bryozoans. Tiny islands unto themselves, the basalt mounds attracted varied marine invertebrates from which a detailed census underscores a web of ecological relationships.[11]

The site's ecology is more nuanced because the low ground among the basalt mounds is densely crowded by large rhodoliths, mostly one-rhodolith deep (fig. 9.10a). Comparisons with the deposits at the Hill of Oranges are apt, and analysis reveals that some of the same algal types are present (*Sporolithon* sp.), but others are different (*Lithophyllum* sp.). A more obvious difference is that the Pedra de Água rhodolith deposit is exceedingly thin. Moreover, the Pedra de Água rhodoliths are larger in diameter and more commonly reveal a large basalt core around which the algae nucleated (fig. 9.10b). Overall, the Pedra de Água rhodoliths exhibit a thin crust compared to those with a rock core at Cabaço das Laranjas (recall Fig. 9.8c). Together with the adjacent basalt mounds, the intervening cover of rhodoliths at Pedra de Água represents an integrated set of ecosystems preserved by catastrophic burial. As at Porto, the overlying strata in the cliff face at Pedra de Água consist of water-laid volcanic ash and lapilli followed by basalt flows. The day's excursion has given us much to ponder as we return to Vila Baleira on Porto Santo. Undoubtedly, the upper island holds still more geological and paleontological secrets to divulge under further scrutiny.

HISTORICAL VIGNETTE:
LIMESTONE MINES ON ILHÉU DE BAIXO

The geographic companion to Ilhúe de Cima off the south shore of Porto Santo is Ilhéu de Baixo (fig. 9.2a), also known by its historic name Ilhéu da Cal, or Lime Island. The older name derives from extensive underground mining conducted for more than three hundred years. Limestone was a strategic commodity used to make mortar for constructing stone buildings both in Porto Santo and Madeira. The industry also focused on manufacturing lime for whitewash used to paint the exteriors of buildings. Limestone shipments from Ilhéu da Cal were exported to Funchal in early 1600 (CE), where the rock was scarce. Limestone continued to be mined during the eighteenth and nineteenth centuries but peaked after the turn of the twentieth century. In 1859, a local tax was levied on the lime business, and the revenue soon became the principal source of municipal income for Porto Santo.

FIGURE 9.10 Miocene shelf at Pedra de Água on Ilhéu de Cima: a) figure seated among basalt knobs encrusted with corals preserved in growth position with feet pointing to a lower level of rhodoliths; b) close-up showing a single layer of rhodoliths with an example of the exposed basalt core around which a rhodolith is encrusted (frame is 20 in or 50 cm on a side).

Photos by author.

The Ilhéu de Baixo mines exploited Miocene strata that outcrop mainly toward the island's south end. The extent of operations is evident on approach from the east any summer evening with a clear sunset when rectangular openings in the cliff face are illuminated by sunlight streaming through the island from the opposite side. Leading to extensive galleries at 165 feet (50 m) above sea level, a claim marker for one of the mines with the letters "BB" is affixed to rocks near the west entrance. These refer to the Blandy Brothers, better known as purveyors of fine Madeira wine. According to family lore, the claim was exploited as a side business during the latter part of the nineteenth century. The roof rock at the BB mine is shored up by square pillars (fig. 9.11), typically a dozen feet (~4 m) on a side, with an average height of 6.5 feet (2 m). It is estimated that as much as 353,000 cubic feet (10,000 m³) of limestone was extracted by hand labor over the mine's working lifetime. Limestone rubble was lowered to the shore and loaded onto boats for direct transport to Funchal, where the family business contracted kilns for the slacking of lime sold as whitewash. Coralline red algae from fractured rhodoliths account for more than 50 percent of identifiable calcareous matter in the mine's product, as shown by analysis of samples retrieved from mine pillars.[12] Indeed, rhodolith-rich limestone was the commercial source of lime processed elsewhere throughout islands in the eastern Atlantic Ocean, including the Azores, Canary, and Cape Verde Archipelagos.

FIGURE 9.11 Outer gallery of the Blandy Brothers' Mine on Ilhéu de Baixo operated for extraction of valuable limestone.

Photo by author.

ASPECTS OF MIOCENE BARNACLE BIOGEOGRAPHY

A detailed Miocene chronology is critical for correlation on a global basis. Dating by nannofossils represented by planktonic unicellular algae is widely consulted in this regard. The pyrogomatid barnacles from Porto Santo, for example, are equated with Calcareous Nanofossil Biozone CN4, also bracketed by volcanic rocks that are between fourteen and fifteen million years old.[13] Discoveries of *Ceratoconcha costata* on Ilhéu de Cima by Santos et al. spurred a resurgence in research on this group and its biogeography.[14] The barnacle and its host corals provide examples of organisms with short-lived reproductive propagules that benefited from oceanic islands as stepping-stones during their trans-Atlantic migrations. In living marine communities, barnacles within the family Pyrgomatidae are among the most common and best-known obligatory associates hosted by shallow-water corals. Studies on such barnacles began with Charles Darwin,[15] who then assigned all coral-hosted barnacles to the genus *Pyrgoma*. Today, the group is split into three subfamilies based on diagnostic features regarding the number of opercular and wall plates possessed. The genus *Ceratoconcha* (subfamily Ceratoconchinae) is distinguished by plain opercular valves and a four-plated wall. It lives exclusively on shallow-water corals largely restricted to the Atlantic Ocean.

Reports on *Ceratoconcha* fossils suggest it appeared almost simultaneously during the early Miocene (twenty million years ago) on opposite sides of the Atlantic Ocean in Florida and the Aquitaine Basin in France. Thereafter, the Middle Miocene Climate Optimum (MMCO) commenced under a warming climate that weakened gyre circulation and promoted stagnation of warm surface water around the equator. As a result, the intertropical convergence zone shifted north of the geographic equator, and coral reefs spread farther north than now. It also seems likely to have been an interval favorable for increased hurricane activity.

The circulation pattern most favorable to long-distance larval dispersal in the Miocene Atlantic Ocean functioned in two directions, with the North Equatorial Current reinforced by convergent trade winds pushing from east to west as opposed to a hurricane-driven pattern from west to east (fig. 9.12). Dispersal was abetted by several Miocene oceanic islands that acted as stepping stones. On the way from east to west, barnacle larvae were likely to hop over the Madeira Tore Rise, the Canary Island Seamount Province, and the Cape Verde Rise, each with exposed islands at that time. Farther west, the Verma and Romanche islands (now

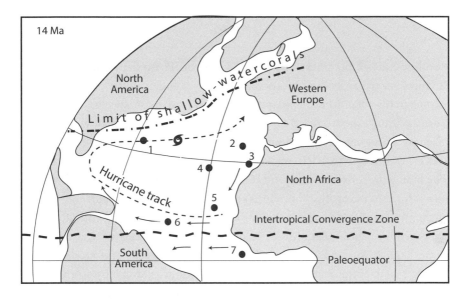

FIGURE 9.12 Palaeogeographic map for the Middle Miocene Atlantic Ocean (modified from Baarli et al. 2017). Position of the Intertropical Conversion Zone (ITCZ) is shown together with the paleoequator and the northern limit for shallow-water corals. Oceanic islands and other features include: 1) Bermuda, 2) Madeira Archipelago, 3) Canary Archipelago, 4) Cruiser Plateau and Plato seamount, 5) Cape Verde Archipelago, 6) Verma tectonic island, and 7) Romanche tectonic island.

submerged) were in place. During the Miocene, Bermuda was a larger island than today, and it sat in the middle of the hurricane track leading from west to east. Moreover, coral-reef hosts reached as far north as Germany and foreshortened the return path of barnacle larvae migration (fig. 9.12).

Reef-building corals declined during the transition between Miocene and Pliocene times and were nearly extirpated from the Mediterranean in southern Europe. Corals and pyrogomatid barnacles disappeared from Madeira. In the Cape Verde Islands, however, fossil corals occur in upper Miocene and Pliocene strata. Surviving coral populations also persisted from the Pleistocene to the present, as found, for example, on Sal Island. *Ceratoconcha* of Pleistocene age at Maio in the Cape Verde Islands represents the first fossil pyrogomatid barnacle recorded from those islands and the last to survive in the North Atlantic Ocean.[16] Thus, the Cape Verde Islands became a refuge for the barnacles and their host corals long after disappearing from the Madeira Islands.

SUMMARY

Localities on the periphery of Ilhéu de Cima off Porto Santo in the Madeira Archipelago tell an ecological story about marine biotas in the coastal zone on a comparably small Miocene island influenced by daily sea swells and episodic storms of hurricane strength. The sheltered shore features a shallow shelf colonized by a fringing reef and mound-dwelling corals augmented by rock oysters and other bivalves that left trace borings both in corals and exposed basalt. The boundary between fossils and underlying basalt is devoid of conglomerate except for the case of algal rhodoliths encrusted around large basalt cores at one of the two localities. Catastrophic burial by water-laid ash and volcanic ejecta preserved biological relationships in place as they once existed on a Miocene coast. The paleoisland's exposed shore exhibits sea stacks and eroded basalt boulders that entrapped thick deposits of rhodoliths swept by storms landward from deeper water. The physical contrast on opposite flanks of the same paleoisland provides a clear window on the difference between fossil communities preserved largely intact and those uprooted from their normal habitat. In this case, Rachel Carson's observations on modern settings are no less prophetic regarding former worlds.

Volcanic islands are naked at birth and await the arrival of life through different means. The principal island of Madeira retains an unusual fossil flora at least two million years old, and the agency of plant dispersal probably entailed the introduction of seeds by birds. Porto Santa's biota from about fifteen million years ago consists of marine invertebrates and coralline red algae, all of which arrived as migrants via reproductive propagules transported from afar by wind-driven currents and storms. The strange example of a pyrogomatid barnacle dependent on cohabitation with shallow-water corals offers the opportunity to more closely examine the Miocene and subsequent history of island-hoping by marine invertebrates with short-lived larval stages.

CHAPTER 10

HOW VOLCANIC ISLANDS RISE, FALL, AND RENEW

A Journey in Early Pliocene Time to the
Azorean Santa Maria Island

The birth of a volcanic island is an event marked by prolonged and violent travail: the forces of the earth striving to create, and the forces of the sea opposing.

—Rachel Carson, *The Sea Around Us* (1951)

Volcanic islands emerge from oceanic depths on a pathway to birth by fire, early death, and potentially short-lived resurrection. Most are doomed to oblivion by erosion, unlike continental islands that survive burial and disinterment to achieve a fixed afterlife in the rock record (chapters 2–8). Classification is nuanced,[1] but many oceanic volcanoes belong to one of three categories. Surtseyan eruptions are hydromagmatic in origin, taking their name from Surtsey, the island that first appeared off Iceland in 1963. That island rose to 430 feet (130 m) before rapid erosion commenced. The extreme violence of Surtseyan volcanoes is due to steam explosions when basalt magma comes in contact with seawater. Growth above the waterline is by an accumulation of fine-grained ash and tephra consisting of air-cooled droplets called lapilli.

Strombolian eruptions are named after Italy's Stromboli Island. Active as recently as October 2022, it has a history of frequent activity stretching back two thousand years. Magma viscosity is intermediate due to bursting gas bubbles once the magma source is isolated from seawater. The resulting stratovolcano grows as a cone with each additional coating of ash, cinders, and volcanic bombs. Hawaiian-style volcanoes are the least explosive due to low-viscosity magma with effusive eruptions leading to shield-shaped edifices. The Big Island of Hawaii is the largest volcanic island in the world, rising 13,680 feet (4,170 m) above sea level.

Hawaiian-style lava flows make the most durable rocky shores out of basalt, which is far more resistant to marine erosion than ash, tephra, or cinder deposits from Surtseyan or Strombolian volcanoes (see chapter 6 on erosion rates). Flows from shield volcanoes behave differently above and below the water line. On land, lava fountains feed lava streams that commonly attain a ropy texture on cooling. This surface results when the upper skin of a flow cools more quickly, while the faster lava current below creates tension against the outer surface to form wrinkles. When a lava lake cools gradually on land from a massive sheet flow, vertical columns form in a polygonal pattern. Pillow basalt is the form most typically acquired by a lava flow extruded underwater. The bulbous appearance is due to a bubble-like expansion of lava, the surface of which cracks open when chilled on all sides by seawater. The crack is the point of issue for the next big pillow extruded from the flow. Basalt of this kind appears as a pile of stony pillows lacking internal bedding planes.

Different volcanic styles and multiple flow patterns accumulate in succession, also interbedded with marine sedimentary layers, to tell the story of an island's rise and fall in reference to sea level. The age of volcanic rocks from different flows may be dated in absolute time based on chemical isotopes undergoing radiometric decay at known rates. Lava deltas that spill into the sea make a predictable sequence that shifts from pillow basalt to columnar basalt as the delta front fills upward from a marine shelf to the water surface and above. In a well-preserved sequence, it is possible to determine the water depth of a former sedimentary seabed by measuring the thickness of overlying pillow basalts to the transition with columnar basalt. This junction in a lava delta is called the passage zone, and it is widely recognized in shore exposures around volcanic islands in the North Atlantic Ocean.[2]

This chapter reviews the Miocene-Pliocene record of emergence, patterns of erosion, and volcanic renewal on Santa Maria Island in the Azores Archipelago of Portugal. The greater island group occurs on both sides of the Mid-Atlantic Ridge (fig. 10.1a). With an area of 37 square miles (97 km²) and a coastal circumference of scarcely 33 miles (53 km), Santa Maria ranks seventh in size among the other islands. It is the oldest on the Azorean plateau and the only one with fossil-bearing strata.[3] From its earliest emergence in the Late Miocene about 6.0 million years ago to the Pleistocene development of marine terraces, the island's geology is most fully treated by Ramalho et al.[4] The Azorean legislature approved certification of the Santa Maria Palaeopoark in 2018.

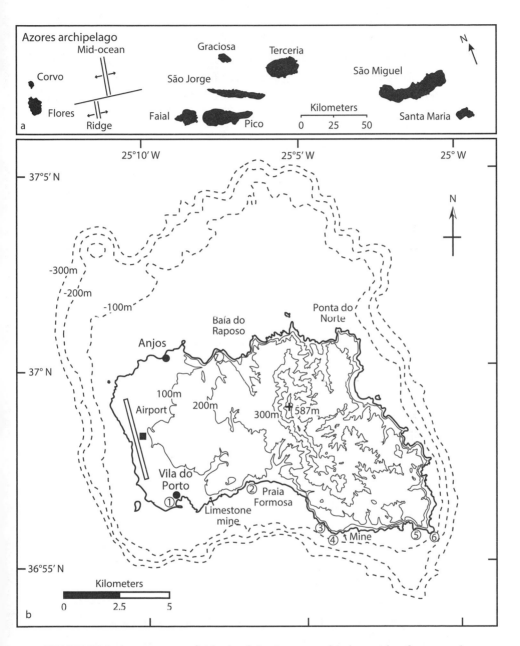

FIGURE 10.1 Azorean maps: a) islands of the Azorean archipelago with reference to the Mid-Atlantic Ocean Ridge; b) combination topographic and bathymetric map for Santa Maria Island (eastern-most island in the Azorean archipelago).

FIRST IMPRESSIONS

A combined topographic and bathymetric map for Santa Maria Island (fig. 10.1b) provides the background for exploring the island's physical geography and attention to its geological history dominated by Pliocene rocks. Rugged rocky shores form roughly 96 percent of Santa Maria's coastline. The airport at the west end is skirted by basaltic cliffs that rise from the sea to 80 feet (~25 m). Elsewhere around the east end, sea cliffs rise more than 330 feet (~100 m). Few beaches exist where a ship's party might easily come ashore, and one marvels at the fortitude of the Portuguese Gonçalo Velho Cabral (1400–1460) and his shipmates who became the first to set foot on the island in 1432 after a sighting on the feast day of Santa Maria. The following year, permanent settlers disembarked on the small cobble beach at Baía dos Anjos (Bay of the Angels). The coastal village and the road that climbs the western plateau are readily visible on landing at the airport. Stone-walled, rectangular paddocks cover the plateau, striking yet another vivid impression. The island is known for beef cattle. Pasturage under a semiarid climate lends the context for calling Santa Maria the Yellow Island.

Santa Maria's highest elevation is Pico Alto in the eastern uplands at 1,925 feet (587 m). The contrast between the open western plateau and forested eastern uplands has an underlying geological basis. A stay of several days in Santa Maria is mandatory for anyone interested in hiking, scuba diving, and experiencing the relationship of geology to the island's overall landscape. A visit to the House of Fossils Museum located on the main thoroughfare of Vila do Porto is a good investment before any such activities. The day trip's focus, described herein, is a boat excursion along the south coast with three stops ashore from west to east to examine aspects unique to the Santa Maria Palaeopark.

FEET ON THE GROUND

Azorean weather is changeable, and it is difficult to predict the day's prospects at daybreak. Despite low clouds and scattered raindrops, the sky clears by 9 A.M. as we gather at the Vila do Porto harbor for an outing aboard an inflatable zodiac. With lunches and water bottles stowed in backpacks, we step off the dock and take seats facing one another along the sides of the craft. The harbor is full of sailing vessels of all sizes, but it also serves as a home port to commercial vessels specializing in tuna fishing at an artisanal scale by the "thrust-and-pull" method using a gaff. Instead of employing nets to take multiple animals all at once, the fish are taken one at a time. The outer harbor includes a ferryboat dock and passenger terminal with its own access road that passes along the base of high cliffs.

In departing Madeira's mid-Miocene islets on Porto Santo (see chapter 9), fewer than ten million years have elapsed. It is the shortest jump in time during our epic journey, and our geologic clock must be precisely reset to commence 5.3 million years ago. The geology in the harbor area is interesting on its own account and may be explored on foot, but the scale is so great that a view from the roadway does little justice to the scene. Not until the zodiac exits the inner harbor and makes a turn through the opening of the outer harbor does the setting become more obvious. Dwarfing the terminal building, a well-shaped volcanic cone of Surtseyan origins reveals a crater summit some 150 feet (~45 m) above the sea surface (locality 1, fig. 10.1b). A cross-section through the structure is outlined by subaerial lava flows that overlap and bury the cone (fig. 10.2). Typical for this time of the year, winds are out of the northeast, which means the south shore is sheltered from ocean swells. The sea surface is barely rippled as the captain opens the throttle and the zodiac picks up speed. Motoring close to shore, it is only to signal the captain that we wish to cut the engine and drift in place when features of interest excite discussion.

Below the windmills at the crest of barren hills, rocky shores reveal a sequence of submarine sheet flows that bracket a thin layer of basaltic conglomerate indicative of emergence close to the Miocene-Pliocene transition. The abandoned Figueiral mine and lime kiln from the late 1800s and early 1900s are out of view in the hillside shrubbery. The facility exploited a layer of Pliocene limestone less than 6.5 feet (2 m) thick, composed mainly of densely packed coralline red algae in the form of rhodoliths.[5] The mine penetrates the hillside for 65 feet (~20 m), leaving pillars in place to support the roof rock. The locality is worth a separate visit on the coastal trail from the village.

Prainha Beach and Fortification. Roughly a mile (1.5 km) farther, we reach a pocket beach, behind which the vegetated hill slope is not as steep as the cliffs we've left behind. Ahead are the ruins of the Prainha Gun Battery, built during the seventeenth century to protect the island from pirates (locality 2, fig. 10.1b). The fortification sits on basement rocks from the Miocene Anjos Volcanic Complex formed by basaltic sheet flows cut by frequent dikes. A contact surface separates those rocks from the overlying Pliocene Touril Volcano-sedimentary Complex with nearly 300 feet (~90 m) of interbedded conglomerate and tuff layers (fig. 10.3), some of which include marine fossils. The basal 80 feet (25 m) features a distinct mixture of conglomerate and breccia in a mud matrix characteristic of a landslide (*lahar*) on the flank of a volcano. When oversteepened,

FIGURE 10.2 Vila do Porto harbor area, showing the cross-section through a Pliocene volcanic cone from a surtseyan-style eruption buried by subaerial lava flows with a car (left center) for scale (see Locality 1, fig. 10.1).

Photo by author.

such landslides may occur under their own weight. The mud mixture, however, implies that debris flows were activated by heavy rainfall. Higher above, two more intervals with conglomerate that incorporates marine fossils show that later debris flows spilled into the ocean. The Pico Alto Volcanic Complex caps the succession (fig. 10.3).

The Pliocene Zanclean stage is represented by an uninterrupted succession of volcano-sedimentary strata in the Touril Complex seated on a major unconformity with the underlying Anjos Volcanic Complex.

The Zanclean stage lasted from 5.3 to 3.6 million years ago.[6] It correlates with widespread rock sequences on Santa Maria Island based on samples from igneous rocks analyzed for radiometric dates below and above the Touril Volcano-sedimentary Complex.[7] A trail with switchbacks gives access to the entire succession and is worth a separate trip accessible from town. We are drawn into Zanclean time (figure 10.3), the start of which is congruent with the start of the Pliocene epoch and the beginning of the Pliocene warm period.[8]

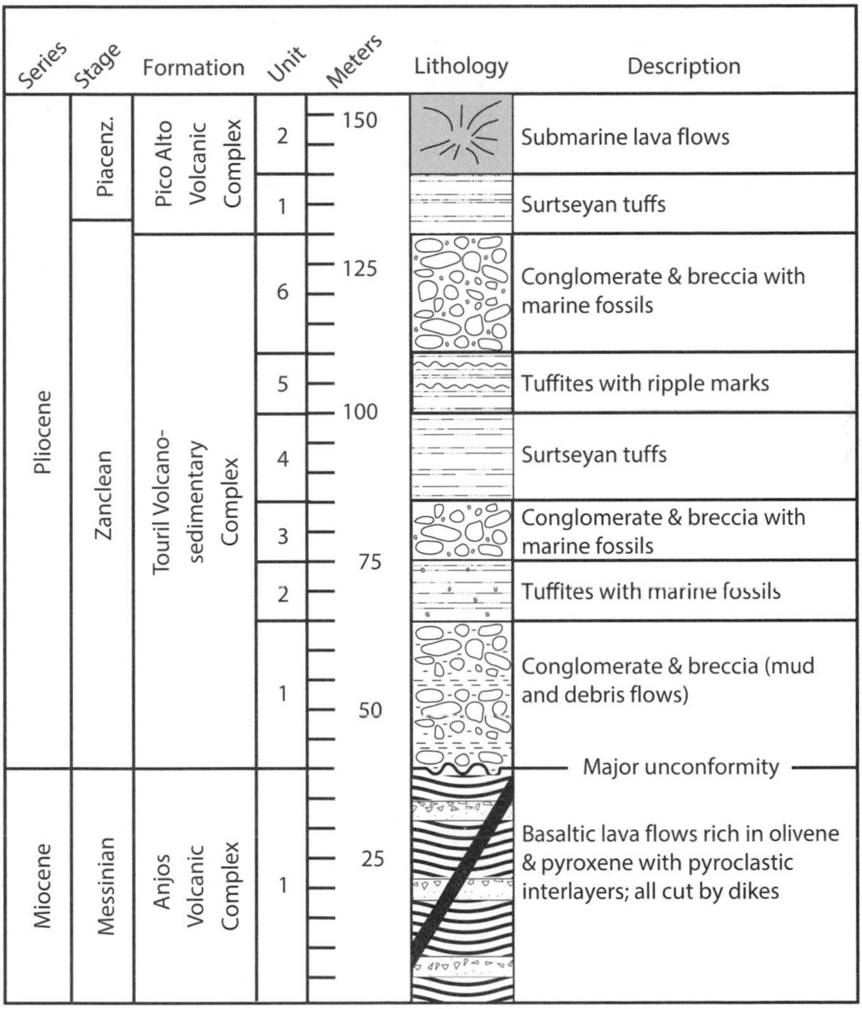

Series	Stage	Formation	Unit	Meters	Lithology	Description
Pliocene	Piacenz.	Pico Alto Volcanic Complex	2	150		Submarine lava flows
			1			Surtseyan tuffs
	Zanclean	Touril Volcano-sedimentary Complex	6	125		Conglomerate & breccia with marine fossils
			5	100		Tuffites with ripple marks
			4			Surtseyan tuffs
			3	75		Conglomerate & breccia with marine fossils
			2			Tuffites with marine fossils
			1	50		Conglomerate & breccia (mud and debris flows)
						— Major unconformity —
Miocene	Messinian	Anjos Volcanic Complex	1	25		Basaltic lava flows rich in olivene & pyroxene with pyroclastic interlayers; all cut by dikes

FIGURE 10.3 Rock sequence in the hillside west of Praia Formosa (see Locality 2, fig. 10.1), based on data collected and shared by Ricardo Ramalho.

Eastward, we cross the bay that fronts Praia Formosa (Beautiful Beach), the island's single greatest stretch of sandy shore. The shelf beneath us slopes gently south for 1.25 miles (2 km) before dropping off more rapidly (fig. 10.1b). The sands of Praia Formosa are washed offshore during stormy weather in the winter season but reappear onshore during the summer to reinstate the most popular swimming beach on the island. A prominent canyon leads through divergent channels to reach an elevation of 650 feet (200 m) about a mile and a half (2.5 km) inland above the beach. A similar canyon occurs on the opposite side of the island at Baía do Raposo (fig. 10.1b). There exists no equivalent beach to Praia Formosa at Baía do Raposa, but the mirrored canyon systems represent the island's most entrenched erosional features. Rock layers from within the canyon were excised by flood events during the last several thousand years. In the distance ahead, it is observed that the Pico Alto Volcanic Complex forms a thicker and more varied sequence than seen previously. A lateral change in layers better fortified by subaerial basalt flows was more resistant to stream erosion than the capstone that protects the underlying tuff and conglomerate layers on the west side of the canyon. Moreover, the contact between the Anjos Volcanic Complex and the overlying Touril Volcano-sedimentary Complex appears to reach closer to the present sea level as we continue eastward.

Ichnofossils' Cave

A mile and a quarter (2 km) beyond the east end of Praia Formosa, we reach our first point of disembarkation at the Ichnofossils' Cave (locality 3, fig. 10.1b). A thin but hard sandstone layer that projects above the cave mouth like an awning forms an overhang some 26 feet (8 m) above the cave floor. It is easily spotted from the sea. The cave is formed by the erosion of thin layers of softer rocks formed by volcanic ash. The seas remain calm, but the landing requires a step upward from the zodiac onto pillow basalts at the shore. It is an extraordinary locality featuring an early Pliocene rocky shore with a slope of welded cobbles at an inclination of 17° that flattens out to a level sandstone floor in the seaward direction (fig. 10.4a). To reimagine the ecology of the place as a former rocky shore, it is necessary to erase all the overlying ash layers, the capping sandstone layer, and all the igneous rocks above that level that belong to the Pico Alto Volcanic Complex. It is an exquisite window into the past that we may close again by reinstating the entombing ash and basalt layers once we depart.

The first order of conduct is to stand back from the cave closer to the present-day shore to take in the physical contours of the setting and the enormity of the original eco-space. More impatient visitors in the background (fig. 10.4a) provide a sense of scale. A person standing on the flat Zanclean sea bed at the foot of the slope would have the top of her head well below the water's surface. That is to say, the water depth between the top of the cobbled slope and the flat seabed amounts to more than 10 feet (~3 m). The next logical order of business is

FIGURE 10.4 Ichnofossil's Cave and selected Zanclean fossils (see Locality 3, fig. 10.1): a) cave overhang, showing gentle slope of a Zanclean rocky shore (upper figures for scale) and level surface of a sandy seabed (lower figures) buried by foreset layers of pryroclastsic sand; b) fossil barnacles preserved in growth position on a basalt cobble at the rocky shore.

Photos by author.

to explore the rocky shore on the cobble slope for signs of the marine organisms that lived there. A close examination reveals abundant fossil barnacles (*Zullobalanus santamariaensis*) preserved in growth position (fig. 10.4b). These were intertidal to shallow subtidal in their preferred habitat. Second in abundance is a large bivalve, *Spondylus concentricus*, preserved with one valve attached to basalt cobbles. Fragments of fossil arc shells (*Arca crassissima*) also occur on the slope, typical for a bivalve that employs a byssus (tough organic strands) to secure a firm hold on a rocky substrate. The platy remains of rock-encrusting coralline red algae are plentiful. Less commonly, examples of small (walnut-size) rhodoliths are scattered among the cobbles.

Moving to the base of the cobbled slope, we find that the level seabed is inter-rupted by small mounds formed by the same cobble substrate that project slightly above the sandstone surface. Around these mini-islands, the remains of large sea urchins (*Clypeaster altus*) lie scattered on the seabed in various orientations. More intriguingly, the sea bed is covered in places by the meandering trackways left by other echinoderms. These are the ichnofossils that make the cave locality so special. Some traces represent burrows made by marine worms (*Dactyloidites ottoi*, and *Macaronichnus segregatis*) and sea urchins (*Bichordites monastiriensis*) before the sand became lithified. Others were worn into the surface of the seabed as shallow trackways made by unknown echinoderms after the sediments were fully consolidated in later Pleistocene time.[9] Unique to the locality, the galleries represent a new genus and species (*Ericichnus bromleyi*) made by sea urchins.

Important clues suggest the extent to which wave activity impacted the Plio-cene rocky shore and adjacent seabed. The first is that the cobbles on the slope have rough surfaces and are angular. They are not the smooth and well-rounded cobbles expected on a shore washed by vigorous wave action. The second obser-vation comes from the seabed with its well-preserved fossil sea urchins. To be sure, the robust and heavy tests typical for species belonging to *Clypeaster* (an extant genus today) are shielded to some extent from damage by strong waves. However, the sandy seabed lacks fragments of broken sea urchins and is largely devoid of crushed shells of any kind. These two factors point to a body of water with a measurable depth of 10 feet (~3 m) that behaved like a sheltered pool subject to little or no wave activity. Exploration in the seaward direction reveals that the pillow basalts on which we landed form a bar partially closing the front of the seabed. In effect, submarine lava that originated in a flow from the west or northwest laid down a protective barrier, much like the artificial harbor con-structed with interlocking concrete blocks back at Vila do Porto. The bar is well exposed at low tide, showing only a narrow opening to the sea. It performs today much as it did four or five million years ago as a barrier against the surf under a rising tide. Sitting in the shade of the cave at lunchtime, we pause to contemplate what it would be like to snorkel in this calm Pliocene pool and witness the full panoply of marine life only hinted from fossil remains.

On departure, the final exercise is to appraise the rest of the geological story in front of us. The paleo-rocky shore, level seabed at its base, and protective bar were buried by repetitive blankets of water-laid sediments formed by vol-canic ash and sand-sized droplets of magma (lapilli). These layers are readily examined in the cove's west wall. Fossil fragments are incorporated within, including recognizable pieces of sea urchins and bits of oyster shells. Truncating the layers is an angular unconformity marked by the sandstone awning over

the present-day cave. Looking carefully, it is possible to spot a beautifully preserved sea urchin (*Clypeaster altus*) half exposed in the cap rock. Above appear
more pillow basalts and the thick succession of lavas belonging to the Pico Alto
Volcanic Complex. As an aside, the ichnofossils' cave was reoccupied as a rocky
shore during the late Pleistocene about 125,000 years ago during the last interglacial rise in sea level.

Malbusca Shoreline

The boat captain waited for us offshore and, at our signal, now returns for us
to continue our tour eastward from the Ichnofossils'cave. A sea arch formed by
basalt lies not far ahead, marking the start of the Malbusca shoreline (locality 4,
fig. 10.1b). Although we will not disembark along this part of the towering
coast, there is much of interest to see and discuss from the boat. Variable but
with a maximum thickness of 80 feet (~25 m), sedimentary layers exposed in
the Malbusca cliffs commence above pillow basalts already 65 feet (20 m) above
the water line (fig. 10.5). The section has been studied intensively by paleontologists to yield a tabulated biota, including seven kinds of marine algae, forty-six
marine invertebrates, and thirteen ichnospecies.[10] At the base of the sequence,
small rhodoliths formed by coralline red algae are preserved in deposits trapped
in surge channels between large basalt pillows. In some cases, the rhodolith
deposits are crossbedded, which indicates they did not live in the channels but
were transported shoreward from an adjacent seabed. Some basalt pillows show
encrustations by barnacles (*Zullobalanus santamariaensis*) and traces of platy
coralline red algae that attest to intertidal conditions within the surge channels.
Shell beds rich in fossil bivalves generally follow above the trapped rhodoliths.
Sandstone, typically crowded with trace fossils of many kinds, follows after the shell
beds. Among the most intriguing trace fossils from this interval are bowl-shaped
depressions left as the feeding signatures of ray fish.[11] Exposed bedding surfaces
at this level feature many such ray holes (fig. 10.6), often surrounded by the traces
left by burrowing sea urchins (*Bichordites* isp.).

Easily recognized from the boat, a massive body of sandstone amounting to
more than 10 feet (~3 m) can be traced laterally as much as 220 yards (~200 m)
along the cliff face (fig. 10.5, lower black arrow). Close up, the sandstone consists
of alternating lamina with black sand and white lime-rich sand packaged in cou-
plets. The black grains derive from finely washed basalt particles. In a few places,
trace fossils penetrate downward from the top of the deposit through sandstone

FIGURE 10.5 Zanclean sedimentary strata in the Touril Complex (layers 1 and 2, black arrows) sitting on pillow basalt (white arrows) at Malbusca (see Locality 4, fig. 10.1). Layer 1 (4-m thick) with uninterrupted sand laminae is a major hurricane deposit, whereas layer 2 represents a succession of lesser storm events.

Photo by author.

otherwise devoid of fossils. Lacking internal subdivisions, the big sand body appears to have been deposited all at once over many hours or perhaps a day, and such a massive feature meets the criteria of a major hurricane deposit.[12] Overlying layers of similar sandstone that amount to nearly 5 feet (1.5 m) exhibit distinct bedding (fig. 10.5, upper black arrow). The upper beds invoke a repetition of lesser storms separated by periods of calm. Given the age of the sedimentary units in this sequence, a correlation fits with the Pliocene warm period when major sea storms and heavy rainfall were thought to be more intense on a global scale under higher average temperatures than today.[13] The sand would have been transported shoreward by storm waves from a large, subtidal sand bar not unlike the present-day bar off the pocket beach at Praínha near the fortification seen earlier in the day.

FIGURE 10.6 Underside of fallen sandstone block at Malbusca with white arrows pointing out feeding traces made by rays (*Piscichnus waitemata*) surrounded by traces from burrowing echinoids (*Bichordites* isp).

Photo by author.

From a closer view of the cliff face (fig. 10.7a), the irregular contact between sedimentary rocks (above) and basalt pillows (below) is emphasized. The distinct break (unconformity) demonstrates well-formed surge channels in cross-section with a maximum relief of 6.5 feet (2 m). Toward the east end of the Malbusca coast, our captain eases the zodiac into a sea cave known as the Gruta de Santa Maria. At low tide, the deep hue of coralline red algae encrusted on pillow basalt makes a vivid impression (fig. 10.7b). Imagining such a scene superimposed on the Pliocene surge channels is not difficult. Notably, the sea cave and the greater Malbusca coast are closer to the edge of the insular shelf than any other place on the island. Sea swells from the south have a short distance to travel before impacting with full force against this part of the shore. A short distance west of the sea cave, we encounter another site of historical interest. At a height above the water, only a little lower than the unconformity shown in fig. 10.7a, there is another limestone mine with a dense concentration of fossil rhodoliths removed, much like the abandoned Figueiral mine. Among the Pliocene fossils recovered from both mines, a large sea conch (*Persististrimbus coronatus*) is significant because it represents the expanded range of a sea snail known elsewhere in the Cape Verde islands as characteristic of tropical conditions.[14]

Pedra-que-Pica

The distance from the sea cave to Pedra-que-Pica is little more than a mile and a half (2.5 km). Direct translation from Portuguese for the name of this site (locality 5, fig. 10.1b) is "stone that stings." The reference alludes to the fact that limestone erosion exposed in the intertidal zone has resulted in a karst-like surface of pointy rocks. The surface is difficult to walk across wearing anything less than sturdy shoes. It requires only a few paces from the landing to reach higher ground out of range from sea splash, where the surface is easier to negotiate. This site has long been the subject of intense study by geologists and paleontologists, and the principal feature on display is a massive deposit dominated by disarticulated bivalve shells that constitute a coquina.[15] *Coquina* is a term explicitly referring to a "death deposit." The original shelly organisms did not live where we now find them but were transported postmortem from another place where they thrived on the insular shelf. Above today's water line, the coquina has a maximum thickness of 13 feet (4 m), but the overall deposit slopes to the east as an inclined shelf. Including the submerged part, it attains a known thickness of 36 feet (11 m). Based on an underwater survey by scuba divers and studies on land, the total area amounts to nearly 6 acres (23,463 m²).

FIGURE 10.7 Exposures on the Malbusca coast of Santa Maria Island: a) Pliocene sedimentary sequence sandwiched between pillow basalt with a highly irregular surface (below) and flat-lying pyroclastic deposits (above); b) nearby sea cave at low tide showing extensive encrustation of basalt by coralline red algae.

All photos by author.

That part of the raised shelf available for inspection on dry land represents a fraction of the known deposit, only about one acre (4,000 m²). However, the deposit extends landward beneath the sea cliffs that rise abruptly to 330 feet (100 m). Hence, the actual extent of the coquina is impossible to gauge. Repeating the same exercise as at ichnofossils' cave, the first task is to ignore the volcanic rocks in the overlying Pico Alto Complex that bury the coquina and imagine the ecological setting as it existed four or five million years ago. Although the nature of the eco-space to the north remains a mystery, we may enjoy an unobstructed view through an arc of 180° from a vantage point with our backs to the sea cliffs. The landing zone where we came ashore is to the south. To the east, we see parts of the submerged shelf reflected by shallow green water as opposed to the dark blue of the deeper ocean. Turning to the west, we view black basaltic rocks on the near horizon standing above the level of the exposed coquina shelf. This is where we must go to continue our immersion in early Pliocene time.

Pillow basalt forms two rocky spurs that project outward to the sea perpendicular to the main shoreline. They represent submarine lava that flowed from a source close by to the north, and they belong to the Anjos Volcanic Complex (see fig. 10.3). These flows may have extended farther out onto the insular shelf, as suggested by a knob of dark rock visible to the south intermittently awash with waves. The more easterly spur rises 30 feet (9 m) above the present sea level and extends uninterrupted for 375 feet (~115 m) beyond our landing place. The spur offers a strategic spot from which to look out across the coquina on its east flank (fig. 10.8a). Erasing the sea cliffs that tower above, the view puts the pillow basalts in the context of a natural wing dam engineered to trap sediments carried by waves and currents arriving along the shore from the east. The spurs were formed underwater, but key clues are found in fractures within the pillows to show that they were uplifted to the intertidal zone and perhaps a little higher before the shell coquina was trapped. Recrossing the easterly spur, a former rocky shore is visible where sedimentary rocks lap against the basalt (fig. 10.8b). The sandstone is a kind of spaghetti rock that incorporates trace fossils (*Macaronichnus segregatis*) left by marine worms regarded as intertidal to very shallow subtidal in their preferred habitat.[16] Sandstone pockets and the fill within fissures are regarded as neptunian dikes, meaning they are sedimentary in origin and not igneous. What is interesting about the dikes is that the "spaghetti" trace fossils occur in life position even within narrow confines to reveal a pattern of upward growth that occurred during a rise in Pliocene sea level. Exposed, now, in cross-section, fissures in large basalt pillows are filled with the "spaghetti" trace fossils that penetrate as much as 13 feet (4 m).

FIGURE 10.8 Pedra-que-pica site and associated Pliocene fossils (see Locality 5, fig. 10.1): a) view east with a rocky spur formed by pillow basalts in the foreground and limestone coquina (black arrows) at center; b) margin of rocky spur (dark basalt) with sandstone containing trace fossils (*Macaronichnus segregatis*) left by marine worms; c) Details of limestone coquina featuring a large pecten (*Gigantopecten latissimus*) and abundant oysters (*Ostrea* sp.).

Photos by author.

The surface of the coquina deposit is packed with fossil oysters (*Ostrea* sp.) and rock oysters (*Spondylus* sp.) in a chaotic jumble of disarticulated and mostly broken shells, among which exceptionally large pectens (*Gigantopecten latissimus*) are strewn preferentially with the outer side of the disarticulated shell facing upward (fig. 10.8c). Commonly up to 4 inches (10 cm) in diameter, only a small number of the shells occur as whole individuals with both valves united. The dominant orientation of such large, saucer-shaped shells with their convex side up is characteristic of a stable position after a turbulent flow of water loses its energy. Whale bones are among the rarest fossils entombed deeper within the deposit. The upper part of the coquina includes rhodoliths formed by coralline red algae and abundant nodules similarly formed by bryozoans. A closer examination of the coquina's upper surface reveals that shells were scoured, and some trace fossils at that level were truncated. Open fissures cut through the coquina to provide a window into the center of the deposit all the way to the underlying pillow basalt where the deposit sits. Two levels of basalt cobbles attest to episodes of erosion from the adjacent basalt spurs. Overall, the scenario conforms to a wave-cut platform on which two or more storm events injected pulses of shelly material swept up from the insular shelf. Given the orientation of the basaltic spurs that constitute both a paleoshore and natural wing dam, the shells were derived from the northeast. The spaghetti trace fossils preserved in growth position at the paleoshore and among the pillow basalts also provide compelling evidence that the entire scenario played out over an interval when the sea level was rising. A succession of storm events probably brought multiple coquinas to the basalt platform only to be washed away by contrary currents. The deposit, now cemented firmly in place, arrived at Pedra-que-Pica when the water depth was well below the level stirred by normal surface waves.

On departure, we are obliged to consider the later Pliocene structures in the Pico Alto Volcanic Complex that overtop the Pliocene coquina and associated paleoshore. A package of volcanic ash and lapilli 118 feet (36 m) thick sits on the scoured coquina, marking an abrupt break in lithology. This material was generated by a nearby Surtseyan volcano with a well-preserved cone, much the same size as that observed in the Vila do Porto harbor (fig. 10.2), also buried in turn by lava flows. At Pedra-que-Pica, a lava delta follows after the pyroclastics, with the passage zone between pillow lavas and subaerial lava flows visible 165 feet (50 m) above the present sea level.[17] It is worth noting that Pedra-que-Pica was reoccupied as a rocky shore during the Late Pleistocene, about 125,000 years ago, during the last interglacial rise in sea level.

FIGURE 10.9 Pliocene storm deposit at Ponta do Castelo (see Locality 6, fig. 10.1b).

Photo by author.

Ponta do Castelo

Rampart-like sea cliffs meet at a turreted corner to impart a well-deserved place name for the southeast end of Santa Maria Island (locality 6, fig. 10.1b). There, a lighthouse with a 46-foot (14-m) tower commands views from nearly 200 feet (~60 m) above sea level. From Pedra-que-Pica, the travel distance is short. A concrete jetty provides a safe landing place well protected from the ocean swell. The site is culturally significant for factory ruins from the island's whale hunting era that remained active until the mid-1960s (see related historical vignette, this chapter). Also found here is a shore exposure of strata 13-feet (4-m) thick belonging to the Touril Volcano-sedimentary Complex.[18] The chief feature on display is a profound erosional break with a sinuous cross-section that cuts deeply into sandstone with alternating lime-rich and ash-rich laminae above coarser-grained sandstone (fig. 10.9). In turn, the troughs are filled by slump beds from 1 to 4 feet (0.3 to 1.5 m) thick formed by medium-grained sandstone. The sandstone includes loose barnacle plates (*Zullobalanus santam-ariaensis*), the spines and complete tests of sea urchins (*Eucidaris tribuloides*), and small sea snails (*Anachis avaroides* and *Alvania sleursi*), all characteristic of an intertidal to shallow subtidal origin. This fauna was carried offshore in a great submarine slide. An overlying lava delta higher in the sequence shows a passage zone 180 feet (55 m) above the present sea level.[19] The implication is that submarine pillows and sheet flows were accommodated in a space of commensurate water depth above the slump beds. In other words, the erosional surface filled by slumped sediments now exposed at the shore was located 180 feet (55 m) below sea level in early Pliocene time.

As at the ichnofossil's cave (locality 3) and Pedra-que-Pica (locality 5), a late Pleistocene shoreline with eroded basalt boulders and intertidal fauna is situated in a thin rim against the volcanic rocks above the slump deposit at Ponta do Castelo. Leaving Ponta do Castelo, the zodiac makes a brief foray around the point along the east coast. Ocean swells carry superimposed whitecaps under a steady breeze from the northeast, and the ride is rough. Our stay is sufficient to take a brief sighting northward across the east coast, where blunt headlands one after another are distinguished by the high-angle strata of lava deltas dipping directly offshore into the sea. At Cedar Point, for example, Lower Pliocene strata characteristic of the Touril Volcano-sedimentary Complex are again partially exposed beneath a cover of volcanic ash.[20] The day's tour, however, is complete, and the zodiac turns back for Vila do Porto.

NORTH SHORE CONSIDERATIONS

Like Ponta do Castilo, two great lava deltas are exposed in the coastal cliffs at Ponta do Norte at the northeast corner of the island. They correlate with the Pico Alto Volcanic Complex, based on reworked fossils within an ash layer.[21] The island's eastern highlands include Pico Alto (fig. 10.1b) and reveal late-stage volcanism with strombolian eruptions and subaerial tuff rings that continued to form while volcanic activity ceased in the west. Pliocene strata rich in fossils could be buried beneath lava deltas, but landfall on the north shore is not possible by boat. Beyond the western headland of Baía do Raposo, however, it is possible to find rhodolith limestone in an abandoned quarry. Interrupted by basalt flows, repetitive volcano-sedimentary layers with marine trace fossils also occur along the high coastal trail east of Anjos. The coarse-grained conglomerate and breccia interbeds on the south shore west of Prahina (locality 2, fig. 10.1b) are unmatched, but the outwash of sediments northward into shallow water is indicated. Comparison of the south and north shores suggests that through Early Pliocene (Zanclean time), Santa Maria Island was a shallow bank covered by marine sediments that slowly subsided while attracting a rich biota of coralline red algae, marine invertebrates, and marine vertebrates, including ray fish and passing whales. During later Pliocene (Piacenzian) time, Surtseyan volcanism was rejuvenated, and the island reappeared above sea level. Whereas volcanism ceased in the western part of the island, activity intensified to the east, resulting in a much thicker accumulation in the Pico Alto Volcanic Complex that included subaerial strombolian eruptions.

SEA STORMS AND REDUCTION OF THE ISLAND PERIMETER

The narrative told by coastal strata on the island's periphery is further amplified by consideration of what lies beneath the sea on the insular shelf (fig. 10.1b). Only twenty thousand years ago, during the maximum expansion of polar ice sheets when moisture was robbed from the oceans and put in cold storage on land, sea level stood almost 400 feet (~120 m) below today's datum. The shelf break around Santa Maria Island falls somewhere between the 650-foot and 1,000-foot (200-m and 300-m) isobaths, where lines of bathymetry are increasingly crowded. During the last glaciation, Santa Maria Island was nearly double in size compared to now. What is most striking is the asymmetry between today's coast and the outer shelf margin compared from north to south. The north shelf is three to four times as wide as the south shelf. Essentially, island size can fluctuate with changes in sea level. When the sea level undergoes a rise during interglacial episodes (as today), coastal erosion has shown a history of being more severe on the north side than elsewhere.

The long-term pattern of sea storms has everything to do with the asymmetry on Santa Maria's insular shelf. Overall, the Azores archipelago is characterized by frequent winter storms that typically arrive from the northwest. Squall after squall, erosion is focused against north-facing shores. During the Atlantic hurricane season, however, the archipelago is in reach of extreme storms once every seven years on average.[22] Major hurricanes that struck the Azores include Hurricane 15 in 1932, Hannah in 1959, Jeanne in 1998, and the two Gordons in 2006 and 2012, respectively. During the last fifty years, hurricane frequency in the Azores has increased with the arrival of shorter but more extreme events.[23]

Rotating counterclockwise, the big storms most often arrive from the south, and the narrow southern shelf on Santa Maria makes that shore more vulnerable. Hurricanes do damage both against the coast with wind-driven waves and inland with excessive rainfall. In the context of the Pliocene warm period, it is no surprise that most of the featured sites on the Santa Maria palaeopark tour include substantial storm deposits.

THE RED DESERT

Above the Baía do Raposo headland (locality 7, fig 10.1b), Barreiro de Faneca is yet another spot of interest. More colloquially, it is known by islanders as the red desert. Covering an area of 20 acres (8.35 hectares), the site was strip-mined for high-quality red clay used to make pottery. On a more industrial scale, the clay was used to manufacture red tiles for the roofs of island houses. The defunct factory where the tiles were baked still exists in the middle of Vila do Porto, marked by a high furnace chimney. The cleared plateau at the red desert exposes the basaltic regolith on which the clay-rich soil developed (fig. 10.10). Rainfall is required to generate such soils, which is more characteristic of tropical settings. The precise age of the clay is hard to establish, but the subaerial basalt flows beneath the clay in this part of the island are regarded as later Pliocene or early-most Pleistocene.[24] This falls outside the Pliocene warm period but still invokes an interval of island history marked by generous rainfall. Similar clearings with red clay are found elsewhere on the island, for example, near Ponta do Norte and on the hilltops of Malbusca (fig. 10.1b).

HISTORICAL VIGNETTE: AZOREAN WHALERS AT PONTE DO CASTELO

Land-based whaling in the Azores began in 1896. The factory at Ponta do Castelo was well organized and continued operations until 1966 when the Companhia Baleeira Mariense went out of business.[25] Among the whaling stations throughout the Azores, the Santa Maria outfit was unusual for its shareholders drawn from a broad base of employees, tradesmen, landowners, and farmhands. The sperm whale (*Physeter microcephalus*), which could grow to an adult male length of 68 feet (~21 m) and weigh 130,000 lbs (59 metric tons), was the sole species

FIGURE 10.10 Cleared land showing weathered red clay and soils at Barreiro do Faneca (also known as the Red Desert) exploited by islanders for pottery and roofing tiles (see Locallity 7, fig. 10.1b)

Photo by author.

FIGURE 10.11 Capstan artifact from the whaling station at Ponta do Castelo used to haul whales ashore for rendering.

———————

Photo by author.

hunted. Oil rendered from the animal's blubber was highly prized as an illuminant in lamps and a machine-tool lubricant. The extra-fine oil from the spermaceti organ was a key ingredient in ointments and cosmetic creams. Whale watchers manned huts on the high south coast at Ponte do Castelo and Malbusca and signaled a sighting by firing a rocket to alert the boat crews. A boat crew consisted of a master helmsman in the stern of a slim rowing boat, the harpooner in the bow who also served as the first oarsman, and five additional oarsmen in between. The boat held a long line to be let out over the bow once the whale was harpooned, and the line was slowly pulled in as the animal tired. The expired animal had to be towed to shore, where a capstan (fig. 10.11) was manually turned to haul the carcass onto the beach.

Processing entailed flensing once the whale's head was cut away. Blubber, which could be a foot thick (~30 cm), was cut into pieces sized to fit into large cast-iron cauldrons in the try-works where the oil was rendered. The opened head allowed access to the spermaceti organ, which might yield as much as 500 gallons (~1,890 liters) of extra-fine oil. In the mid-1950s, the Ponte do Castelo factory was upgraded with a more efficient autoclave for heating the cauldrons. The intricate machinery sits rusted in the try-works. At its peak, the company employed a crew of forty who operated four whaleboats and two motor launches to help with the recovery of the whales. After thirty years of activity, the Companhia Baleeira Mariense concluded with a capture tally of 850 sperm whales, yielding 4,400 pounds (2,000 kg) of processed oil. Pliocene whale bones from Pedra-que-Pica testify to the millions of years that whales have migrated through the Azores during the spring and summer months.

PLIOCENE ISLANDS ELSEWHERE

Mexico's Gulf of California includes many well-preserved paleoislands represented by Lower Pliocene (Zanclean) limestone and sandstone that crown tilted layers of Miocene andesite. The surface flows of volcanic andesite, chemically intermediate between basalt and rhyolite with a SiO_2 content of around 60 percent, preceded the gulf's initial flooding by a few million years. When the Baja California peninsula pulled away from the Mexican mainland, the stretched crust led to block-fault islands distributed all along the 685-mile (1,100-km) arm off the Pacific Ocean. Punta Chivato is located some 330 miles (530 km) due south of the U.S.-Mexican border, covering an area of only 9.5 square miles (25 km²) in the shape of a cross-piece (*atravesada*) pointed like an elbow straight into the Gulf of California. This district is defined geologically by a cluster of paleoislands called the Santa Inés Archipelago after a pair of member islands still surrounded by seawater. A network of informal trails from the center of the area loop out in all directions to provide full access, with detailed descriptions of the local geology.[26]

Although the outer radius of the five paleoislands at Punta Chivato is small compared to Pliocene Santa Maria in the Azores, the monadnocks are the key to unraveling environmental conditions that affected all the other block-fault paleoislands in the Gulf of California. The most pervasive relationship is the influence of a winter wind (*El Norte*) that still exists today due to a wind field driven by high-pressure cells funneled from north to south down the axis of the gulf. Paleontologically, there is strong evidence for windward (north-facing) and leeward (south-facing) habitats that contrast with one another around the Santa Inés paleoislands. Island shores open to the north feature clean limestone with dome-shaped coral colonies (most commonly *Solenastrea fairbanksi*), whereas sheltered shores open to the south are dominated by large oysters (*Ostrea* sp.) preserved in a silty limestone. Fossil pectens comprise other facies found on ramps that slope in an easterly direction. All three variations contribute to strata that form a continuous skirt around the paleoislands.

Especially noteworthy is the coral species endemic to the Gulf of California (*Solenastrea fairbanksi*) but with nearest relatives in the same genus from the Caribbean Sea. It is believed that the ancestors of the gulf corals migrated westward from the Caribbean through the straits of Panama prior to 3.5 million years ago when the Isthmus of Panama linked North and South America to form a barrier against further marine interchanges. Similarly, Pliocene bryozoans that encrusted andesite cobbles in a basin south of Punta Chivato represent the same species (*Conopeum commensale*) still living on the Atlantic Coast of North America.[27] These represent the only bryozoans of their kind to colonize an outpost on the Pacific Ocean. It is a repetition of the same pattern told by Cretaceous rudist clams (see chapter 8), which managed to establish a foothold on the North American West Coast after migrating westward through the Tethys Seaway.

GLOBAL PLIOCENE GEOGRAPHY

Early Pliocene (Zanclean) geography in place five million years ago is not much different than today. However, a small detail with drastic implications occurs in the open passage at the Panama straits that allowed a strong current to pass from the North Atlantic Ocean to the North Pacific Ocean (fig. 10.12). Evidence for an open passageway prior to 3.5 million years ago comes from Pliocene corals and bryozoans in the Gulf of California linked to the Caribbean Sea and North Atlantic Ocean. A useful comparison may be made with modern-day currents that flow through the Indonesian Archipelago along the equator to set up a bulge of water in the eastern Indian Ocean, driving the eccentric north-to-south Leeuwin Current off the coast of Western Australia.[28] The south Indian Ocean gyre remains strong but is displaced farther offshore Western Australia to accommodate the Leeuwin Current. It is inferred that a vigorous, through-flowing current at the straits of Panama would induce comparable

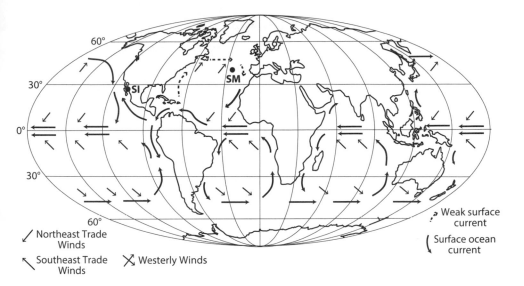

FIGURE 10.12 Global reconstruction for the Early Pliocene (Zanclean) world with interpretations for trade winds convergent on the paleoequator, high-latitude westerly winds, and surface ocean circulation. SM = Santa Maria Island in the North Atlantic Ocean; SI = Santa Ines archipelago in the Gulf of California (Mexico).

countercurrents flowing north along the shores of Central America to the Gulf of California and south along the shores of Ecuador and Peru.

The expansion of warm tropical water in the eastern Pacific Ocean was abetted by the countercurrents resulting from the through-flow of seawater at the straits of Panama. Such a scenario with the pooling of warm water is credited as fostering persistent El Niño conditions that brought a substantial increase in storms and rainfall to the West Coast of North America.[29] Another consequence of an open passage between North and South America would be a weakening of the Gulf Stream Current that flows north along the eastern seaboard of North America before crossing from Newfoundland to the British Isles. The northeast trade winds may have become less effective, which would result in a similar pooling of warm seawater. Under this scenario, the steerage of hurricanes over the North Atlantic Ocean was altered enough to send storms with a higher frequency toward the Azores. Based on the discovery of the large sea conch (*Persististrombus coronatus*) from the Lower Pliocene of Santa Maria, it is reasonable to conclude that mean annual sea surface temperature (SST) was from 6.5°F to >11°F (~3.7°C to 6.3°C) higher than the current average SST of 69° (20.6 C).[30]

SUMMARY

Formed before other Azorean islands, Santa Maria was a lone outpost in the North Atlantic Ocean. It rose from the sea, sank back again during an interval of extensive erosion when the Touril Volcano-sedimentary Complex was deposited, and experienced rebirth from a combination of uplift and volcanic activity that broke the surface with Surtseyan and Strombolian vigor. Rachel Carson's prose is a fitting statement in this example of the struggle between volcanic sources making new land and the sea's erosional response. During the intermediate stage of planation, the flat-topped edifice of Santa Maria was barely awash with seawater and attracted a diverse biota, including marine algae, marine invertebrates, and marine vertebrates represented by abundant ray fish and migrating whales.

Paleontological and geological evidence points to a surprisingly tropical setting for a place fixed well above the Tropic of Cancer at a latitude of 37°N. Warmer surface ocean water is implied by certain fossils, and the warmer water became a magnet for heightened storm activity. Normally outside the track of present-day hurricanes, most localities described on the island's south shore feature Lower Pliocene (Zanclean stage) strata indicative of major storms. Ichnofossils' Cave (locality 3, fig. 10.1b) is the exception, interpreted as a sheltered rocky shore with a quiet pool-like basin behind a protective bar. The narrowness of the insular shelf in the south shows that the Pliocene and present-day coastlines are closely aligned. They were and remain vulnerable to hurricanes. Over time, more frequent winter squalls have eaten away at the north shore, resulting in a wider insular shelf. Red-clay deposits at Barreiro da Faneca and elsewhere on the island were formed as soils before the close of the Pliocene but clearly after the global Pliocene warm period. Even so, the clays point to an annual cycle of wet and dry conditions long preceding the island's present climate.

CHAPTER 11

HOW THE YOUNGEST ISLANDS CHALLENGE WITNESS

Journeys in Pleistocene Time to Islands on the African and Pacific Tectonic Plates

> *Sometimes the shore speaks of the earth and its own creation; sometimes it speaks of life. If we are lucky in choosing our time and place, we may witness a spectacle that echoes vast and elemental things.*
>
> —Rachel Carson, *Our Ever-Changing Shore* (1957)

Planet Earth's outer crust is fractured by several large tectonic plates and an equal number of smaller plates, each with dynamic and ever-changing boundaries. The plates undergo processes of three main types. In the first, two plates may diverge from one another, spreading apart at oceanic ridges where basalt magma rises to fill submarine rift valleys from below. In the second type, the oceanic edge of one plate may converge against the land margin of another at trenches, where dense slabs of basalt plow beneath more buoyant granite. The third type of activity entails collisions at the boundary between the oceanic margins of two convergent plates or between the continental margins of two such plates. The former leads to volcanic action resulting in island arcs like Japan, whereas the latter brings about massive mountain ranges like the Himalayas of Nepal and Bhutan caught in a squeeze between India and Tibet.

The plate-tectonics revolution in the earth sciences erupted during the mid-1960s at a time when the career of Rachel Carson jumped from oceanography and the related coastal-marine sciences to contend with the issue of pesticide contamination on land. Although cognizant of the Atlantic midocean ridge, she had no clue about its importance as a place where new ocean crust is made and spreads apart with the divergence of tectonic plates. Even so, her intuition was prescient on the fathomless source of inspiration stimulated by physical and biological variations found on shorelines worldwide. The sinuous path of the Mid-Atlantic Ridge between North America and Europe, as well as South America and Africa, was charted for the first time well after the twenty-two-year-old Charles Darwin left Portsmouth, England, to accompany the mapping expedition of the HMS *Beagle* (1831–1836) under Captain Robert FitzRoy. A few years later, the twenty-five-year-old James Dwight Dana departed Hampton Roads, Virginia, assigned to the USNS *Peacock* on the United States Exploring Expedition (1838–1842) under Lieutenant Charles Wilkes. Darwin and Dana seized the time and circumstances of their island travels to advance bold theories relevant in unexpected ways to concepts of plate tectonics and island biogeography yet to come. The canvas of plate tectonics is truly global in scale, but the scientific observations of two nineteenth-century naturalists are well-framed within the boundaries of two very different tectonic plates. Indeed, their contributions echo today as early probes into Carson's sense of "vast and elemental things."

TALE OF TWO TECTONIC PLATES

None of the tectonic plates that divide Earth's surface contrast more than the African and Pacific Plates. The African Plate (fig. 11.1) extends far out from the center of a continent, where midocean ridges close around like a castle moat on three sides through the Atlantic, Southern (Antarctic), and Indian Oceans. As the ridges denote zones of crustal expansion, they show that the African Plate has a history of steady enlargement. The north margin is a convergent boundary against Europe that passes through the Mediterranean Sea with a consequential history of disturbance associated with the Swiss and Carnic Alps as well as active volcanism at Mount Etna in Sicily and the isle of Stromboli (see chapter 10). Regarding freeboard on its surrounding oceans, Africa is unique among all other continents with the highest average land elevation. This is not the same as possessing the world's highest mountains but relates instead to Africa's general buoyancy like an empty supertanker riding high in the water. In part, such extreme buoyancy portends the future breakup of the continent along the East Africa Rift traced by interconnected lakes, including Victoria, Albert, and Tanganyika. The area of ocean crust surrounding the continent that counts as part of the African Plate is equal to the land area of the continent itself. Numerous oceanic islands dot this broad belt. Some were visited and later described by Darwin,

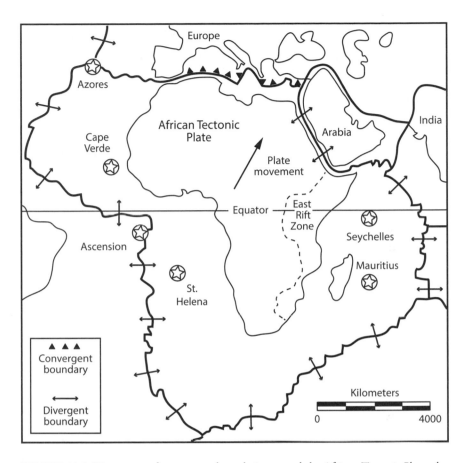

FIGURE 11.1 Divergent and convergent boundaries around the African Tectonic Plate, also showing some oceanic islands visited by Charles Darwin during the 1831 to 1836 voyage of the H.M.S. *Beagle*. The Seychelles Islands is an exception, known to Darwin (1842) but not on the basis of personal experience.

including Mauritius, St. Helena, Ascension, the Cape Verde Islands, and the Azores (fig. 11.1). Other prominent groups include the Madeira (see chapter 9), Canary, São Tomé and Principe, Comoros, Réunion and Seychelles Archipelagos, and the big island of Madagascar.

The Pacific Plate lacks a core continent. However, it is half-defined over an immense distance by a divergent boundary called the East Pacific Rise (fig. 11.2). Like the Mid-Atlantic Ridge, it represents a zone where basalt magma floods submarine rift valleys on a regular basis. Unlike the Mid-Atlantic Ridge, its orientation in the Pacific Basin is not midocean. In fact, the East Pacific Rise reaches into the Gulf of California between the Baja California peninsula and

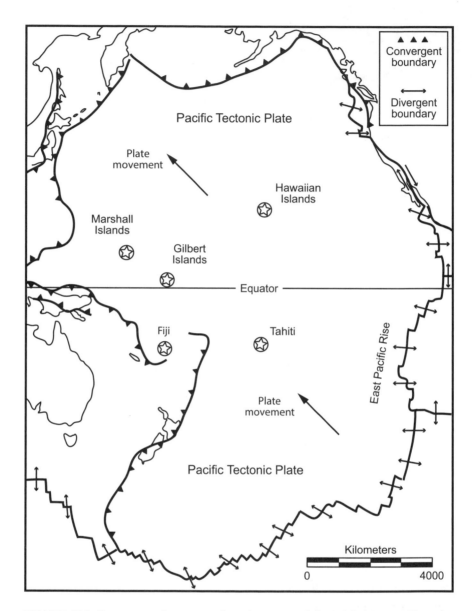

FIGURE 11.2 Divergent and convergent boundaries around the Pacific Tectonic Plate, also showing oceanic islands visited by James Dwight Dana during the 1838 to 1842 United States Exploring Expedition. Tahiti was the only overlap with Darwin's earlier circumnavigation of the globe.

mainland Mexico, where it merges with the San Andreas fault to transfer a slice of North America to the Pacific Plate (fig. 11.2). A roughly equal length is traced by convergent boundaries where the Pacific Plate is pushed (or perhaps pulled by the density of its basalt slabs) beneath the Aleutian Islands of Alaska, Japan, the Philippines, New Guinea, and parts of New Zealand. Charles Darwin stopped in Tahiti late in his voyage on the HMS *Beagle*, and that island represents the only point of overlap with the more extensive exploration of Pacific Plate islands investigated by James Dwight Dana from the *Peacock*. Although the United States Exploring Expedition transited outbound through the North and South Atlantic Oceans, its work was focused on the Pacific Basin. Islands in the hundreds are scattered across the Pacific Plate, and Dana visited and described many. His direct experience included not only Tahiti, but also Fiji, the Gilbert Islands, the Marshall Islands, and most significantly, the Hawaiian Islands (fig. 11.2).

The physical geography of Pacific Plate islands and their attendant marine biology illustrate a radical departure from African Plate islands, particularly those on the Atlantic seaboard. Pacific islands are dominated by coral reefs, whereas the West African islands are without coral reefs. Beginning more than two-a-half million years ago, the Pleistocene history of these islands conveys a consistent story. Darwin and Dana were instrumental in teasing out many of the geological and paleontological details of that story. In so doing, they raised more questions than were answered. In hindsight, we can appreciate that some variations are influenced by different styles of plate-tectonic action, whereas others remain as much a mystery today as they were for the voyagers on the *Beagle* and the *Peacock*.

GROUNDED ON ISLANDS ASTRIDE THE AFRICAN TECTONIC PLATE

Even the smallest of islands seated on the largest of tectonic plates bear witness to details regarding present and former conditions of marine circulation and biology that are strongly influenced by regional patterns. Islands from among two archipelagos on opposite sides of the Africa Plate are avatars for most island clusters surrounding the continent. Santiago Island belongs to the Cape Verde Archipelago on Africa's Atlantic seaboard, and it played a crucial role in the maturity of the young Charles Darwin as a geologist. La Digue Island is part of the Seychelles Archipelago on Africa's opposite seaboard. Although he never visited the archipelago, Darwin was familiar with it through research conducted soon after he returned home following his circumnavigation of the globe aboard the *Beagle*. Oahu belongs to the Hawaiian Archipelago, one of the most isolated island groups in the North Pacific Ocean. Dana spent much time exploring the geology of that island during repeat visits of the *Peacock*. His insights influenced observations made on neighboring islands in the Hawaiian chain and others in the Marshall and Gilbert Islands.

Praia District of Santiago Island

Santiago (St. Jago to Darwin) is the largest island in the Cape Verde Archipelago, located 370 miles (~600 km) off the West African coast (figs. 11.3a, b). Flights from Boston to Praia, the capital city of the Cape Verde Republic, require little more than six hours of flying time. It took three weeks under sail for the *Beagle* to reach Praia harbor leaving Portsmouth on December 27, 1831. Captain FitzRoy was eager to find a suitable place to calibrate the ship's chronometers needed to fix calculations for longitude vital to his mapping mission. The harbor islet of Santa Maria (Quail Island to Darwin) below the Praia plateau became the staging ground for that exercise (fig. 11.3c, locality 1). Including a return visit near the conclusion of the *Beagle*'s voyage, Darwin spent almost a month exploring the geology within a three-mile (5 km) radius of what was then a Portuguese colonial outpost.

The tripart layering of a 6-foot (1.8-m) thick limestone seam sandwiched between basalt flows on Darwin's Quail Island can be viewed from the south side of the Old Town plateau and from the boulevard leading to the harbor area directly across from the islet. From the ferry terminal at the harbor, Rua da Achada Grande climbs an escarpment crossing through the same sequence (fig. 11.3c, locality 2). An outcrop bordering the road on the left exhibits conglomerate with large basalt cobbles on which the limestone sits, overlain by basalt. Here, the limestone's texture consists of finely broken rhodoliths. The few macrofossils include casts of large sea snails (*Persististrombus coronatus*), characteristic of a shallow-water setting adjacent to a rocky paleoshore. Farther up the hill, an outcrop exhibits the wrinkled surface texture of a subaerial basalt flow, commonly called ropey lava because of the coil-like strands that form during cooling (fig. 11.4a). Returning to the bottom of the hill, we may enter the harbor area, where a parallel road climbs the same escarpment to reach warehouses on the flats above.

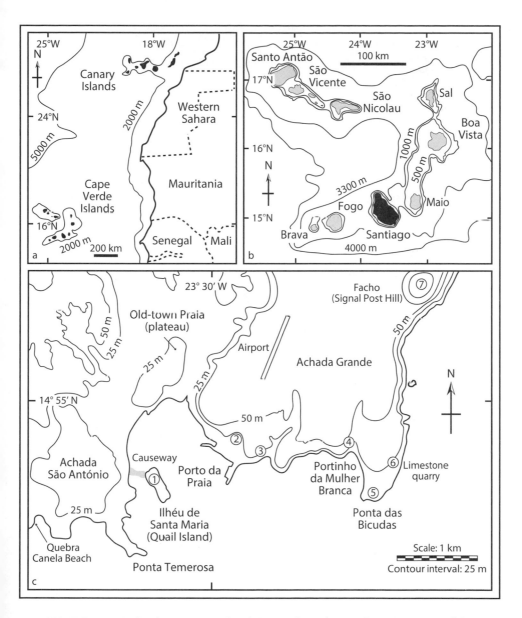

FIGURE 11.3 Archipelagos associated with West Africa: a) maps showing position of the Canary and Cape Verde Islands; b) principal islands of the Cape Verde archipelago with emphasis on Santiago (black); c) topographic map covering the southeast part of Santiago Island, including the town of Praia, capitol of the Cape Verde Republic.

FIGURE 11.4 Pleistocene outcrops in the Praia harbor area: a) road cut at the side of Rua da Achada Grande showing ropey lava characteristic of a subaerial basalt flow (hammer for scale); b) road cut at the side of the parallel harbor road showing limestone dominantly composed of rhodoliths overlain by pillow basalt (white lines define the toe of a lava delta).

Photos by author.

The outcrop on the harbor road exposes limestone with rhodoliths the size of golf balls (fig. 11.4b). Darwin correctly identified the dominant component of the limestone as an accumulation of concretions precipitated by "nulliporae," or coralline red algae.[1] Fossils collected on Quail Island are archived in the Sedgwick Museum at the University of Cambridge, leaving no doubt that the budding naturalist was among the first to recognize fossil rhodoliths. The term *nulliporae* is no longer applied by those who study marine algae today, but the descriptive meaning "without pores" clearly differentiates these plants from coral colonies that preserve small pits occupied in life by individual animals with stinging tentacles. Darwin further observed that many of the algal concretions are nucleated around basalt pebbles. He inferred that spherical growth must have occurred under conditions during which the rhodoliths were in near-constant motion subject to wave activity in relatively shallow water.

Examples of pillow basalt follow above the limestone in the long section on the harbor road (fig. 11.4b). Here, clumps of solidified magma are commonly 1 foot (~30 cm) in diameter. The term *pillow basalt* was not part of Darwin's vocabulary, but he understood the correlation between such rocks and the flow of molten lava into seawater, and he expressed a fascination with the chemical reaction of lava in direct contact with lime-rich sediments. Moreover, the harbor section is instructive because pillows are found at the toe of a lava delta descending from above (see white borders, fig. 11.4b). The thickness of pillow basalt within the lava delta amounts to 40 feet (12 m), as found in the measured road cut.[2] Above this level, lava that cooled under subaerial conditions takes over. The numbers provide a kind of dip-stick to indicate that a former seabed with abundant rhodoliths was no more than 40 feet (12 m) deep when a stream of lava spilled into the ocean from a nearby source on land.

We follow along the top of sea cliffs east from the harbor road for two-thirds of a mile (1 km) to reach a path descending into a ravine for the next stop at Portinho da Mulher Branca (fig. 11.3, locality 4). The inlet where fishing dories often are beached translates from the Portuguese as the "Little Port of the White Woman." Climbing through talus accumulated at the base of the sea cliffs, sufficient height is attained for a spectacular view of the natural coastline stretching back to the harbor area. Contacts both below and above the limestone reveal key differences. The upper contact with the overlying pillow basalt is horizontal, just as it appeared in the road cut (fig. 11.4b).

In contrast, the lower contact between basement basalt and the succeeding limestone is uneven, undulating with pronounced highs and lows. Darwin correctly interpreted the scene as a former rocky shoreline where the lows correspond to the intersection of drainages from the land, not unlike the ravine from which we just arrived. As the sea level rose, rhodoliths and sediments enriched by rhodolith sand filled the low places before overriding the upper parts. After some time, the super-abundant rhodoliths formed a level sea bed. In turn, the seabed was catastrophically impacted by a lava flood that spilled offshore through coalescing deltas. In quick succession, individual flows piled up underwater to eventually reach the surface. Subsequent flows were land-based from start to finish, providing the raw material for a seaward extension of the coastline. We see before us the same scene that Charles Darwin first saw on a January day in 1832.

A sand bar crosses the mouth of the ravine, giving easy access to the headland of Ponta das Bicudas (Barracuda Point) to the southeast (fig. 11.3c, locality 5). This, too, is ground on which the young Darwin walked. At low tide, a network of tidal pools spreads along a low rocky shore where isolated blocks of basalt provide shade. The shadows are a welcome relief from the bright sunlight, and we are tempted to open our lunch packs before noon. Coral colonies of a single species (*Siderastrea radians*) are barely covered by seawater at low tide. Their mustard-yellow coloration contrasts sharply with patches of black basalt left open here and there. Sea urchins with intermediate spines are the principal occupants free to move around the tide pools. A devoted beetle collector, it was likely at Ponta das Bicudas that the young Englishman encountered his first corals at the start of what would become a six-year voyage. The scene left a deep impression, which Darwin recalled forty-four years later when writing his biography. "How distinctly I can call to mind the low cliff of lava beneath which I rested, with sun glaring hot, a few strange desert plants growing near, and with living corals in tidal pools at my feet."[3] It was at that moment, so he claimed, when the young Darwin felt the rising ambition to author a book on the geology of volcanic islands that he was certain to visit during the ensuing voyage.

Based on his detailed cross-section that traces basalt and limestone layers along the coastline to the north, it is clear that Darwin knew the area quite well. Our exploration resumes to the northeast to reach a small limestone quarry cut into the cliffs close to shore (fig. 11.3c, locality 6). Surviving tool marks show it was a working quarry where valuable limestone was extracted, probably for the manufacture of lime. Darwin does not mention this spot, and it's most likely that the quarry operation commenced sometime after the *Beagle* sailed from Santiago for the last time in 1836. As described by Baarli et al., the quarry floor rises step-wise over basalt basement rocks through 8 feet (2.5 m) and inland over 32 feet (~10 m).[4] Given Darwin's excitement with the nearby tidal pools populated by coral colonies, he would have been ecstatic by the discovery of comparable corals preserved as fossils in the quarry assemblage. The extraction of limestone was not total, leaving contacts intact between an irregular basalt surface and overlying limestone at multiple levels.

The lowest quarry level is perched on the edge of a sea cliff that rises 16 feet (~5 m) from a turbulent sea below, fully exposed to the trade winds. Atop the cliff, basalt knobs protrude through limestone formed by rhodolith debris (fig. 11.5a).

FIGURE 11.5 Quarry locality north of Ponta das Bicudas: a) limestone shelf 16 ft (~5 m) above sea level, showing protruding knobs of basalt (hammer for scale); b) fossil coral (*Siderastrea radians*) on basalt; c) fossil fire coral (*Millepora alcicornis*) on basalt.

All photos by author.

Many knobs are occupied by corals of the same species (*Siderastrea radians*) living in the tide pools at Ponta das Bicudas. Here, the fossil corals are robust with thick growth (fig. 11.5b). The uppermost part of the quarry exhibits encrustations left by the fire coral (*Millepora alcicornis*), actually a hydrozoan rather than a true anthozoan coral (fig. 11.5c). Pediments of broken spires remain where the hydrozoan grew upward from a mat. In life, the colony would appear bright red or orange, as its living counterpart does today. The same zone at this level includes large barnacles (*Megabalanus azoricus*), likewise attached to basalt. Quarry levels in between reveal oysters (*Ostrea* sp. and *Spondylus* sp.) cemented to basalt in life position. Likewise, examples of a small, solitary cup coral (*Balanophyllia* sp.) are attached to basalt. Depressions in the basalt became the last resting place of cone shells (*Conus* sp.). Bivalves commonly called turkey wings (*Arca* sp.) are lodged within cracks in the basalt. Adding to the brocade of ecological details, several different trace fossils may be tracked through the carbonate sand, including varieties with complicated feeding patterns (*Rhizocorallium jenense versum*).[5]

Most of the fossils from the quarry limestone are represented by organisms still found in parts of the Cape Verde Islands at water depths from 13 to 33 feet (4 to 10 m). Considering the quarry site's elevation, it is apparent that Santiago's southeast coast underwent an uplift of no less than 70 feet (21 m). Darwin was well aware of the circumstances related to this phenomenon, as he was intent on tracing the same rock layers along the coast between Quail Island and Signal Post Hill (fig. 11.3c). He was convinced these strata were "Tertiary" in age. Based on the radiometric dating of basalt in Praia harbor, lava flows that buried the limestone date from 700,000 years ago, with an accuracy of plus or minus 200,000 years.[6] Hence, coastal uplift postdates that time and clearly falls later than the age assessment postulated by Darwin.

Northward from the quarry, the limestone shelf narrows to a mere pathway. Here, fossil rhodoliths the size of golf balls are abundant, and splendid examples of pillow basalt are found in cross-section in the adjacent cliff face directly on the fossils. Darwin identified marble only once in his descriptions of the rock layers around Praia, and that spot was surely here. The rhodoliths have a hardness and luster akin to porcelain. One can almost feel the transfer of heat to the limestone from the superincumbent pillows that look as freshly formed as yesterday. A short distance beyond the quarry, we cross the first cove eroded in the steep rocky shore and reach thereafter a second. Fishermen frequent this spot to cast lines from the rocks. A trail leads straight upward to the top of the cliffs. Nothing is within sight to betray our presence in the twenty-first century as we scramble after Darwin to reach the top of the plateau nearly 100 feet (30 m) above the sea.

Our goal is the volcanic crater at Facho, known to Darwin as Signal Post Hill (fig. 11.3c, locality 7). The crater's circular lip rises to 460 feet (140 m) above sea level, and the steepest part of the ascent remains on crossing the flats of the Achada Grande. Darwin described the crater, some 525 feet (160 m) in diameter, in close detail with observations on the merger of lava flows onto the plateau below. Descending from the outside of the crater on its northeast slope to continue along the coast, we finally arrive at a spot where the seam of Pleistocene limestone appears to pinch out in the sea cliffs below. The relationship is one that Darwin illustrated by means of a carefully drafted cross-section.[7] Although he gives no explanation for the cutout, it is clear that Facho was one of several volcanoes that erupted from radial fractures during a final phase of activity, adding more basalt to the island's circumference. Much has changed since Darwin's time, not least of which is the addition of a busy airport. Commercial aircraft that dwarf the HMS *Beagle* in size arrive several times a week on flights from Boston and Lisbon. A regular grid of paved streets sprawls across the flats below Facho to the southwest, and we follow them to where a fleet of minivans operate on an irregular schedule to move people back and forth from the town center in Praia.

La Digue in the Seychelles Islands

From the opposite side of Africa (fig. 11.1), the Seychelles in the Indian Ocean provides a marked contrast to the Cape Verdes. Notably, the sixteen inner islands of the archipelago are formed by granite. Smaller or larger, the Seychelles Islands are mountaintops on a granite plain covering 517,377 square miles (1,340,000 km²), with an average water depth between 145 and 215 feet (~45 and 65 m). Several types of granite are recognized throughout the islands, but pink granite is dominant on La Digue and nearby Prasline Islands, whereas gray granite on Mahé is more coarsely grained. Overall, the geologic age falls within an interval around 750 million years ago during the Late Proterozoic, with affinities to similar granites on the Horn of Africa.[8] In short, the Seychelles Islands are a splinter of Africa that broke off and drifted away but remain as part of the African Plate.

Neither Cape Verde nor the Seychelles Archipelagos were lost in the sense of geological burials that put them out of sight. However, the two island groups are sufficiently distant from the African mainland that they were unpeopled until quite late in human history. The French claimed possession of the Seychelles in 1756, long after the Portuguese settled the Cape Verdes in 1462. The name derives from Vicomte Moreau de Seychelle, who was the

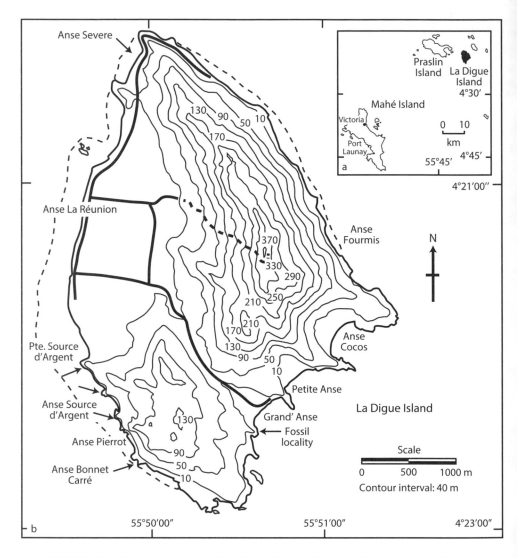

FIGURE 11.6 Topographic map of La Digue Island in the Seychelles archipelago: a) inner islands of the Seychelles with La Digue profiled in black; b) enlarged map of La Digue showing named bays and headlands.

controller-general of France at that time. Permanent settlers arrived in 1770, making the largest island of Mahé (fig. 11.6a) a center of commerce. Two years earlier, a company of Frenchmen arrived on the vessel *Digue* with orders to harvest timber for shipbuilding. The name La Digue was affixed to the island where their activities occurred.

Only 4 square miles (~10 km²) in area, La Digue ranks as the fourth-largest island in the Seychelles group. Its coastal circumference is 9.5 miles (~15.5 km), all but the extreme southern end of which is accessible by roads and pathways (fig. 11.6b). Belle Vue is the highest elevation at 1,214 feet (370 m), giving the island a prominence of 1,320 feet (~400 m) above the surrounding submarine shelf. Among island monadnocks, La Digue is only a quarter the size of Porto Santo in the Madeira Archipelago (see chapter 9) and barely a tenth the size of Santa Maria in the Azores (see chapter 10). But it is 45 percent larger than South Korea's Hongdo (see chapter 1). By composition, however, La Digue is fully granitic, and the island features the best exposures of fossil-rich limestone in direct contact with granite anywhere in the Seychelles.[9] For that matter, it might be claimed that no better place exists than the Seychelles to examine the relationship between granite rocky shores and limestone.

The only way to reach La Digue is by ferryboat from Victoria on Mahé. Once on the island, bicycles may be rented. There are some twenty named beaches on La Digue, amounting to a cumulative frontage nearly half the island's circumference. From the community center at Anse La Réunion, we may follow a road through a narrow valley separating the northern two-thirds of the island from the rest to reach the shores at Grand'Anse (fig. 11.6b). Leaving the bicycles on the beach, it is a short walk to a fossil locality on the granite headland at the south end of the beach. The sea cliffs here are wonderfully sculptured (fig. 11.7a), rising abruptly as much as 60 feet (~18 m) above the shore. Vertical granite walls are fluted, becoming knobby in profile at the top. Locally, the limestone occurs in unexpected forms. An exposure only an inch thick (2.5 cm) is plastered like stucco against the pink granite (center, fig. 11.7a). It covers a surface area of two square yards (1.75 m²) but was far more extensive when waves last washed this spot some 125,000 years ago during the Late Pleistocene. A line of horizontal crenulations can be seen from below on the underside of a massive granite ledge. Most are filled with limestone that adheres to the roof rock. The width of each segment is 40 inches (16 cm) across, completely filled with limestone deposited in thin sheets that faithfully trace the shape of the overlying arch (fig. 11.7b). These sheets were deposited by coralline red algae (*Lithophyllum* sp.), one of the types that also forms freerolling rhodoliths. The deposit has left a high-water mark, attesting to a former rise in sea level.

A pocket of limestone rich in the fossils of marine invertebrates crops out a little farther south in the cliffs at the same level as the crenulations with the platy limestone. The fossils from this site include dome-shaped corals (*Favia favus*) that sit directly on granite.[10] There are limpets (*Cellana radiate*), specialized for grazing on rocky surfaces. Several kinds of fossil sea snails are in evidence at

FIGURE 11.7 Granite coast and fossil coralline red algae on La Dique Island: a) fluted sculpture eroded in granite at Grand'Anse (person for scale). Photo by B. Gudveig Baarli; b) thinly layered coralline red algae cemented to roof flutes of overlying granite (ruler for scale in inches).

Photos by B. Gudveig Baarli (a) and author (b).

this locality (*Nerita plicata* and *N. undulata*; *Morula granulate*; *Conus frigidus*; and *Trochus virgatus*). All have descendants that thrive today on intertidal flats around the Seychelles Islands. The limestone matrix enclosing the fossils also includes quartz grains eroded from the adjoining granite that testify to physical erosion on a rocky shore exposed to the southeast trade winds.

The coast between Pte. Source d' Argent and Anse Bonnet Carré on the opposite side of the island (fig. 11.5b) is equally attractive for explorations that bring visitors to small beaches separated by sculptured granite headlands. Pockets of limestone are hidden all along this part of the coast. As at Grand'Anse, the pink granite exhibits adhering sheets of calcareous algae. In places, preservation is clear enough to show the surface texture of the fossil algae.[11] Scattered as they are, limestone coatings show that the weathered granite surfaces so bizarre in shape existed unchanged since the late Pleistocene.

HISTORICAL VIGNETTE: JAMES DWIGHT DANA IN THE HAWAIIAN ISLANDS

The Hawaiian Archipelago includes eight major islands, sensu stricto, arranged in a curvilinear pattern, with the Big Island of Hawaii in the southeast and the smaller island of Kauai at the opposite end 350 miles (~565 km) to the northwest. The Big Island has remained volcanically active, with a history of continuous lava flows since 1983. In a broader sense, the island chain extends northwest from Kauai another 550 miles (885 km) to Midway Island, with numerous smaller islands and atolls in between.

Before Hawaii became a territorial protectorate in 1898 and finally gained American statehood in 1959, the islands drew scientific attention during the United States Exploring Expedition of 1838 to 1842. Among the participating academics, none were more luminous than James Dwight Dana in terms of thorough studies on island geology and coral biology.[12] The USNS *Peacock*, to which Dana was assigned, was stationed in the Hawaiian Islands from September 24 to December 3, 1840. He spent five days on the Big Island, where he trekked across the Mauna Loa and Kilauea shield volcanoes. Due to Dana's influence, these remain the archetype examples for the oceanic shield volcanos' form and eruptive behavior. Eruptive styles subsidiary to central calderas that result in cinder cones and tuff cones along fissures (now called rift arms) were noted by Dana.[13]

On Oahu, Dana toured raised coral reef formations around the island's cir-cumference, now attributed to the Pleistocene Waimanalo Formation. The most extensive occur along the southern and southwestern shores, where he recorded variations in thickness between 15 and 30 feet (4.5 and 9 m). Dana interpreted island uplift to have been on the order of 30 feet (9 m). Extensive growth rep-resented by fossil corals was noted on the island's western side near Pupukea, as were additional reef deposits to the northeast at Kahuku Point and at Laie Beach and Waimanalo Beach (fig. 11.8) to the southeast. Dana also attributed the presence of former beaches and dunes on the windward side of the island to the accumulation of coral-eroded sand. Dune rocks with large-scale cross beds are especially well exposed at Laie Point.

After touring the Gilbert and Marshall Archipelagos, the *Peacock* returned to Honolulu harbor for repairs from June 14 to 21, 1842. This gave Dana addi-tional time to continue with investigations on Oahu. Sometime during his two early visits to Hawaii, Dana landed on Kauai, where he took stock of that island's physical geography and state of erosion. In contrast to other islands

FIGURE 11.8 Pleistocene coral-reef overlain by bedded marine sand on the southeast side of Oahu, near Waimanalo Beach.

Photo by: Daniel R. Muhs.

in the chain, Dana concluded that Kauai was not necessarily older, but that volcanism had long since become quiescent. Less than a decade apart, Dana and Darwin witnessed different oceanic volcanoes in the Pacific and Atlantic Basins, which led to contrasting ways in which they perceived the development of those islands.

APPRAISALS OF ISLAND PHYSICAL GEOGRAPHY

The volume on volcanic islands envisioned by Charles Darwin as early as 1832 while still on Santiago eventually came to fruition in 1844 but was preceded in 1842 by a different book that earned him a reputation among geologists after the HMS *Beagle*'s homecoming. The 1842 contribution set out the hypothesis of atoll formation from preexisting volcanoes that undergo subsidence while surrounding coral reefs continue to grow upward at the same pace.[14] Some islands in the Pacific Basin have fringing reefs around a rocky coastline, many exhibit barrier reefs with a moat of shallow water between a volcanic island and the surrounding reef, whereas more than a few are represented by a circle of sandy land barely above sea level distinguished by a shallow interior lagoon and exterior coral reef. After his initial encounter with living corals on Santiago, Darwin was transfixed by the sight of luxuriant coral reefs when the *Beagle* reached Tahiti on November 15, 1835. His only experience on an island atoll occurred when the *Beagle* reached the Cocos (Keeling) Islands of the Indian Ocean in April 1836. Based on those experiences, Darwin undertook a comprehensive research project on coral islands that led to the 1842 volume. The USNS *Peacock*, with James D. Dana aboard, occupied the same Tahiti anchorage only three years after the *Beagle* on September 12, 1838. Later in Sydney, Australia, Dana read an account summarizing Darwin's concept of atoll formation.

When Darwin's attention returned to the larger pattern of island origins that led to his 1844 volume, he saw the physical geography of archipelagoes "either in single, double, or triple rows, in lines which are frequently curved in a slight degree" and where "each separate island is either rounded or more generally elongated in the same direction with the group in which it stands." Contrary orientations in the Galapagos, Canary, and Cape Verde islands were cited by Darwin while admitting those in the Cape Verdes are the "least symmetrical of any oceanic, volcanic archipelago." In his 1849 volume, Dana exhibited a sophisticated view of volcanoes in "two parallel series" of alignment for the Hawaiian Islands, as accepted today. He also understood that the

Hawaiian chain extended farther northwest beyond Kauai to other islets and atolls. However, Dana mistakenly argued that the big island of Hawaii was not the youngest in the group but "only the last of the number to become extinct." He believed all Hawaiian Islands had a simultaneous origin, fed by magma from the same parallel fractures that closed progressively in one direction from west to east. The same reasoning was applied to the linear chains of volcanic islands for which he had direct knowledge, such as the Marshall and Gilbert Archipelagoes (fig. 11.2). Simultaneously, Dana clearly applied criteria based on the physical erosion and subsidence of islands starting from pristine volcanic shields and ending with the morphology of Darwin-class atolls.

As forward-thinking as Darwin and Dana were, they reflect no preconception of plate tectonics. All oceanic islands visited first-hand by the two naturalists are intraplate volcanos related to stationary mantle plumes (figs. 11.1 and 11.2). Darwin's intimacy with islands on the periphery of the African Plate was radically different from Dana's island chains on the Pacific Plate. The Pacific Plate moves to the northwest at an average speed of four inches (10 cm) a year, and the island volcanoes it embodies are relatively short-lived because they shift away from the source of magma and submit thereafter to erosion. The smallest islands are the farthest removed from the magma source. Most of the islands on the African Plate are related to divergent boundaries that expand at a slower rate—around a half-inch (2 cm) per year, on average. They, too, are linked to mantle plumes but make longer-lasting volcanoes buoyed by ongoing injections of magma to island clusters that experience episodic uplift. An exception to the pattern is the Madeira Archipelago (see chapter 9), which is related to a chain of seamounts aligned to the northeast in the same direction as the overall movement by the African Plate (fig. 11.1). The uncharacteristic signs of island uplift so well documented by Dana on Oahu are, today, attributed to crustal flexure at a limited distance from the Big Island of Hawaii that subjects the plate to a heavy mass.[15]

APPRAISALS OF ISLAND BIOGEOGRAPHY

In the vernacular of island biography, islands are divided into categories depending on how large or small they are and how far they are from the nearest mainland.[16] The Hawaiian Islands, for example, vary in size from the Big Island at 4,028 square miles (10,432 km²) to Ni'ihau at 70 square miles (180 km²). In principle, the larger the island, the greater the number of species it can support. Regardless of size, all Hawaiian Islands are isolated from the nearest continental mainland, 1,860 miles (~3,000 km) away. Each one had zero species when it broke the surface of the ocean in volcanic violence. The assembly rules that govern how many species might thrive on or around any given island and which species might contribute to a dominant group are complicated. It is difficult for marine and terrestrial biotas alike to find their way to the Hawaiian Archipelago. Marine organisms have a natural advantage over terrestrial organisms because their reproductive propagules can be transported over long distances by marine currents (see also chapter 8 on the Cretaceous). Even so, the number of coral reef species and reef fish that populate the Hawaiian Islands today is quite small (450 species) compared to the Marshall Islands (1,000 species) closer to the equator within the east-west equatorial current.

In addition to his work as a geologist, Dana was a pioneer in the study of coral reef species,[17] adding solid contributions to our understanding of the physical limitations that control the spread of coral species by larval dispersal. Better remembered by students today for his mineralogy textbook, the Yale professor was an adept biologist who documented to what extent the geography of coral reefs is divided into latitudinal districts: the torrid (or tropical) and subtorrid (or subtropical) zones. According to Dana's experience formulated in his 1872 volume, peak coral diversity appeared in the Fiji Islands, where the "temperature of the surface is never below 74° F (23°C) for any month of the year and all the prominent genera of reef-forming species are abundantly represented." In contrast, Dana found that "The Hawaiian Islands, in the north Pacific, between the latitudes 19° and 22° are outside the torrid zone of oceanic temperature, in the subtorrid, and the corals are consequently less luxuriant and much fewer in species." Dana also observed that Hawaiian waters support a profusion of "hardier" species in the dominant genera of finger corals (*Porites* sp.) and stag-horn corals (*Pocillipora* sp.). It all came down to which species in larval form might travel the farthest from centers of dispersal and tolerate the coolest surface water.

Darwin's principal expertise in marine zoology was with barnacles in the phylum Arthropoda (see chapter 9). Barnacles are expressed by male and female members. As they are stationary individuals in adulthood, migration is possible only during the early larval stage of development following fertilization. One of Darwin's claims to zoological fame was his investigation of reproduction in goose-neck barnacles (*Scalpellum scalpellum*) in which a dwarf male bonds with a normal-size female protected within the female's shelly carapace.[18] When fertilization occurs under close quarters, it is a safe ecological investment. More commonly, male barnacles release their propagules into the water, and fertilization occurs at a high ecological cost. Moreover, barnacle larvae tend to be short-lived before metamorphosis takes hold, and the maturing animal must find a hard surface for attachment. Hence, free-swimming barnacle larvae cannot hitch a ride with an ocean current for any prolonged length of time. There are exceptions, such as when barnacles colonize pumice rafts that result from volcanic eruptions and the floating rocks drift with ocean currents. Several generations of barnacles may live attached to such rafts before they wash ashore on a distant island. Others are specialized in their attachment to whales that move around the oceans on migratory paths.

Terrestrial plant migration from one island to another is another matter. Here, too, Darwin played a role, conducting innovative experiments in his backyard at Down House in Kent. Enlisting his children as helpers, Darwin subjected vegetation to saltwater and retrieved seed samples to check for germination at regular intervals.[19] A wide selection of plants was provided from Kew Gardens by his friend, Joseph Hooker. The object was to determine how long the seeds for any particular species might tolerate saltwater before losing vitality. Darwin concluded that 10 percent of the seed species tested under his experimental regime could float for thirty days, and assuming the speed of an ocean current at 33 miles (53 km) per day, those might readily cross an ocean to germinate in foreign soil. Coconut palms come to mind as the fabled trees of tropical isles capable of long-distance migration by a floating seed pod that is almost indestructible. The classic case of island isolation leading to an endemic species is the Coco de Mer (*Lodoicea maldivica*) from the Seychelles Islands. Otherwise, an ordinary-looking palm tree, the female tree (fig. 11.9a) is unique in producing the world's heaviest seed, weighing as much as 48.5 lbs (22 kg). The female flower is the largest of any palm tree. The male tree is unique for its enormous catkins (fig. 11.9b), up to 3 feet (~1 m) in length. Today, the species is largely restricted to Praslin Island in the Seychelles, although seed pods are reported to wash ashore as far away as the Maldives Archipelago.

CONTRASTING PLEISTOCENE AND OLDER LIMESTONE DEPOSITS

Santiago, near the middle of the Cape Verde Archipelago, and Oahu, at the center of the Hawaiian Archipelago, are endowed with rich deposits of Pleistocene limestone. Each of these islands, situated far apart and on their respective tectonic plates, preserve deposits that could not be more different. Oahu is encircled by living coral reefs, and its fossil reefs represent an interval when sea level was higher than today. Santiago has no protective barrier reefs. Excluding

FIGURE 11.9 Coco de Mer palms (*Lodoicea maldivica*) in the Valle de Mai Nature Reserve on Praslin Island: a) female tree (arrow pointing to nut); b) male tree with elongated catkins (arrows).

Photos by author.

the coral buildups at Ponta das Bicudas (Figure 11.5), the Pleistocene limestone of Santiago is dominated by algal rhodoliths. Indeed, the limestone archives of island groups throughout the northeast Atlantic, including the Azores, Canary, and Cape Verde Archipelagos, are dominated by rhodoliths.[20] Except for the Miocene of Madeira (see chapter 9), the oceanic shelf of West Africa has remained without substantial coral reefs for most of its geologic history. From direct experience, Charles Darwin understood the overall absence of coral reefs in the North and South Atlantic oceans, and he raised the possibility that river sediment flowing into the Gulf of Guinea may have hindered coral growth. In his 1842 study on coral reefs, Darwin argued: "But the islands of St. Helena, Ascension, the Cape Verde, St. Paul's, and Fernando Noronha are, also, entirely without reefs, although they lie far out at sea, are composed of the same ancient volcanic rocks, and have the same general form, with those islands in the Pacific, the shores of which are surrounded by gigantic walls of coral-rock." For his part, James D. Dana became the acknowledged master of nineteenth-century science in consideration of Pacific Ocean islands sheltered by fringing and barrier coral reefs. The Pleistocene record on those islands clearly reflects an affinity for reef development. Moreover, various drilling programs on atolls during the late nineteenth to middle twentieth centuries not only confirmed Darwin's 1842 hypothesis of atoll development but also substantiated the lengthy geologic history of coral reefs in the Pacific and Indian Oceans (see more in chapter 12).

SUMMARY

Separated from one another by the North and South American Plates, the African and Pacific Plates comprise a combined surface area amounting to roughly 38 percent of Planet Earth's outer crust. No two tectonic plates could be more different—one centered around a continent with a high freeboard in excess of any other continent and the other lacking a continental core. Save for the Seychelles, the islands that populate the two plates reflect geophysical constraints that regulate the rise and fall of volcanic islands in response to faster or slower plate movements over fixed hotspots unrelated to plate boundaries. As early as 1844, Darwin attempted to rationalize the symmetry of nonlinear island groups off the West African coast and emphasized Pleistocene limestone as proof of widespread island uplift. Darwin's acclaim as a geologist arrived early in 1842 with his formulation of the atoll hypothesis to explain widespread island subsidence in the Pacific and Indian Oceans. By 1849, Dana achieved success as a geologist by combining Darwin's atoll model with a sophisticated analysis of age-based island geomorphology. He realized that the relative age of islands in linear chains resulted from progressive stages of surface erosion and subsidence but adhered to the mistaken notion that volcanism began at the same time along any one line of islands. Darwin was the first to appreciate rhodoliths as substantial contributors to limestone, but he never came to understand just how pervasive these organic concretions actually are around islands in the North and South Atlantic Oceans. Why that is so remains unresolved. The mysteries of "vast and elemental things" in coastal settings alluded to by Rachel Carson are far from depleted.

CHAPTER 12

HOW ISLANDS DRAW MEANING AND OBLIGATION

Descending Mount Misen on Japan's Sacred Miyajima

The pleasures, the values of contact with the natural world, are not reserved for the scientists. They are available to anyone who will place himself under the influence of a lonely mountain top—or the sea.

—Rachel Carson, *Lost Woods* (1954)

In their geologic context, islands are woven into the fabric of deep time traced through hundreds of millions of years. Islands possess no actual meaning in that framework, being objects of no moral inclination. Ancient as they are, islands record a physical presence, each with a lifespan having a fixed birth and death. However brief that interval may have been in each manifestation, biological life almost always secured a foothold on island shores. Hence, the story of island life is embroidered through deep time with the unfolding of evolution. Only through the human experience do islands acquire shades of meaning and are put to many different uses. What, then, are we to think about islands today and tomorrow and the significance they portend for humankind's stewardship of Planet Earth? How is it that as late arrivals to this sphere, we humans find so many different ways to use and abuse islands?

Geologists constitute a distinct mindset within a larger collective of scientists. The norms within that larger guild exert an expectation for impartiality and respect for the facts of the physical and biological sciences, where ever they may lead. As implied by Rachel Carson, scientists would not be doing their work without the simple gratification earned in fulfilling those tasks. After an eventful career bringing me to many distant and unexpected corners of the world, this geologist finds himself conscience-bound to express veneration for our home planet and the thread of life it has fostered, born of its immense longevity. Taking the word at face value, as invoked by Rachel Carsen in *Silent Spring*, how can we continence the growing *pollution* of our only home in the vast universe? I am not alone among those who affirm that something akin to a higher faith may serve as a motivation to use the geological sciences for the development of a sustainable future.[1]

Islands are among the most fragile ecosystems in our world due to vulnerabilities inherent in their small size compared to continents. Loss of biodiversity appears sooner and more drastically in island settings than on the nearest mainland. The concluding chapter of this narrative journey aims to catalog the uses and abuses to which we subject islands. Planet Earth is a mere speck of an island in relation to our solar system, the Milky Way, and the rest of the universe. Is it too late to suggest that the most positive attributes we humans assign to islands may help us to restore to good health our life raft in the universe, whereas other traits appeal to our darker instincts, whether intentional or not?

ISLANDS AS SACRED PLACES

The starting point for our journey across geologic time began with the riddle of Mount Monadnock in southern New Hampshire. Not a genuine island in the sense of a watery setting, nonetheless, Mount Monadnock is the prime exemplar of a sky island. Indeed, Mount Monadnock lends its status to all sky islands and related seascapes with drowned monadnocks the world over.[2] On continents subject to large-scale changes in relative sea level over time, many paleoislands that once were genuine islands have since reverted to sky islands. Thus, the distinction between sky and aqueous islands is somewhat interchangeable. The dozen paleoislands featured in chapters 2–11 are shown in their present-day distribution on a global map, also marking tectonic spreading zones (fig. 12.1). The sacred islands summarized in this segment are plotted on the same map. As it happens, Mount Monadnock is at the geographic center of the American transcendentalist movement made famous by the poet-philosopher Ralph Waldo Emerson in the mid-nineteenth century with other luminaries like Henry David Thoreau. Meaning is drawn from Mount Monadnock in the sense that an all-embracing essence is said to radiate jointly from raw nature and from human beings who are a part of nature. In their links with people, several of the paleoislands described in earlier chapters are regarded as holy places in their own right, irrespective

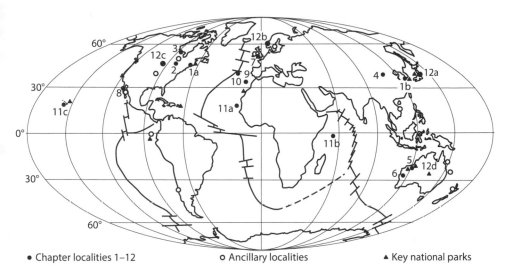

FIGURE 12.1 Global distribution of all chapter localities (black circles), ancillary localities (open circles), and key national parks (black triangles) discussed in this book. Also depicted are the major spreading centers where tectonic plates are actively pushing apart.

of their scientific identity as drowned monadnocks. The fact is that hallowed ground derives from different but related sources of human intuition.

The endpoint for our journey is deliberately chosen with an acknowledgment of holiness and a philosophical link to the transcendentalist movement unforeseen in the ninth century. It is Miyajima on Japan's Seto Inland Sea, crowned by Mount Misen. The island is treasured for its twelfth-century floating shrine, but temples with an even older history are near the top of the mountain. The most renowned is the Reikado, or Hall of the Spiritual Flame. The building shelters the eternal flame said by tradition to have been kindled by Kukai in 806 CE, the monk better known by his honorific title Kobo Daishi.[3] As a twenty-eight-year-old student, the young monk went to China on a state-sponsored mission to learn esoteric Buddhist teachings, which he did in the Tang dynastic capital of Xi'an. On his return to Japan, the Kobo Daishi founded the Shingon sect of Buddhism, unique for its teachings that spiritual bliss (nirvana) is communicated by natural phenomena in all their variations and may be achieved in one's lifetime without passing through multiple stages of rebirth. Kukai also won distinction in 821 CE as the civil engineer who restored the great Manno Reservoir on Shikoku, Japan's fourth largest island. The lake reservoir remains one of the largest of its kind in Japan, still in use as a source of irrigation water. Folklore posits that Kukai worked tirelessly for the good of farmers by locating many springs throughout Shikoku as sources for lesser irrigation projects. Today, the island is home to eighty-eight Shingon temples, which attract pilgrims on the spiritual quest to visit all eighty-eight shrines traditionally on foot and dressed in white robes. Done properly, it is a perambulation that may take several months to accomplish.[4]

Miyajima is a granite island with a prominence of 1,755 feet (535 m). A trail system gives access to the heart of the island through a circular loop, although a tramway makes the ascent easy. Descent may be taken by the Omoto Route (about 2.8 miles or 4.5 km), which passes the cave where Kukai is reputed to have meditated during his one-hundred-day residency on the mountain. The formative experience for the visitor is leaving the island following immersion in nature when all must pass the Great Tori (fig. 12.2) on their return to society. It is by way of our human interactions that the opportunity exists to make a difference in daily life, for better or worse. The Kobo Daishi made such a difference, setting a living example held in high esteem many centuries afterward.[5]

Others, influenced by island nature but from distinctly different backgrounds, left their mark on the social world they inhabited. Petter Dass (1647–1707) accrued a larger-than-life reputation as the parish priest from Alsten Island on the Norwegian Sea near the Arctic Circle. The district's church at Alstahaug sits at the base of a granite mountain range called the Seven Sisters. Rooted in one of the most scenic and awe-inspiring parts of coastal Norway, the Seven Sisters stand apart from one another, separated by hanging valleys smoothed by the flow of glacial ice. The narrative poem written by Dass, *The Trumpet of Nordland* (1739), not only describes the Seven Sisters of his homeland (see following section on island mythology) but other distinctive features in the coastal landscape.[6] As one of the so-called potato priests of that era, Dass not only tended to the spiritual needs of his flock but also sought ways to materially improve their welfare in a place of severe physical hardships at the mercy of short summers and long winter seasons.

The granite monadnock at the center of the Black Hills in South Dakota is Black Elk Peak, named for Nicholas Black Elk (1863–1950), the sage of the Oglala Sioux, who drew spiritual strength from that spot. Black Elk reflected on the significance of the site: "There, when I was young, the spirts took me in my vision to the center of the earth and showed me all the good things in the sacred hoop of the world."[7] As we have seen (chapter 2), the Black Hills are on the flank of a larger paleoisland mapped as Siouxia on the transcontinental arch of Cambrian North America by paleontologist Christina Lochman-Balk.[8] With the ebb and flow of continental seas during the last quarter of the Cambrian period, Black Elk's monadnock cycled back and forth between a genuine marine

FIGURE 12.2 Great Tori Gate at the entrance to the Itsukushima Shrine on Miyajima in Japan's Seto Inland Sea.

Photo by author.

island surrounded by a shallow sea and a sky island encircled by a wide plain. Later during the Cretaceous period, the continental seas returned, and the Black Hills were surrounded by seawater once again. Today, Black Elk Peak and the rest of the Black Hills stand out as a prominent sky island on the Great Plains with a distinctive natural history.[9]

Likewise, the quartzite prominence that became a marine island during the latter part of the Ordovician period (see chapter 3) is revered to the Ojibwe clan of the White River First Nation in Ontario, Canada. Dreamer's Rock on Birch

Island is close by the highway that links the rest of Ontario with Manitoulin Island in the Great Lakes. It is a protected site on tribal land for which permission to visit must be granted by tribal authorities.[10] Bater Obo in China's Inner Mongolia (see chapter 4 on the Silurian system) is devoted to the personality of Genghis Khan, who swept across the steppes of Asia as a military commander during the early thirteenth century. Dedication of this particular mountaintop as a memorial to a historical figure is an appropriation of the broader worship afforded the Mongolian god of "Eternal Heaven," known as Tengri. A geographic high point, or obo in Mongolian expression, is a physical spot nearer to heaven and, thus, worthy of Tengri's acknowledgment.

In keeping with a wider spiritual link to nature, the aboriginal peoples of Australia honor every geologic prominence, stream crossing, and other natural phenomena across the width and length of the land (see chapter 5 on the Devonian system). None are more iconic than the towering monolith at Uluru (Ayer's Rock) near the center of the island continent (fig. 12.3a and b). The site is part of the Uluru-Kata Tjuta National Park (fig. 12.1), under supervision by Indigenous Australians. In contrast to Africa (chapter 11), it is worth noting that Australia demonstrates the lowest freeboard of any continent on the planet. Unlike the empty supertanker of the African continent that rides high above the water line, Australia behaves like a heavily loaded vessel with its gunwales nearly awash by the sea. Time and time again, Australia has been flooded by continental seas that covered much of the territory. In this regard, it is possible to reflect on Uluru as one of those monadnocks that alternates through geologic history between a sky island and something closer to a watery island.

ISLANDS IN STORY AND MYTH

Islands are revered in story and myth in ways other than those with a strictly spiritual connection. Examples are myriad, but two with strong symbolism are worthy of mention. Long handed down through oral storytelling, Homer's *Odyssey* ranks among the earliest adventure sagas on record. Odysseus (also known by his Latin name: Ulysses) was the legendary king of Ithaca who came up with the idea of the Trojan horse that led to the vanquish of Troy in Asia Minor. Following the sack of that city, it took Odysseus ten years to return home to his island kingdom. Almost all of the places visited by Odysseus and his men during their years of hardship are tied by folklore to real Mediterranean islands. Ithaca is a small Greek island (45.5 square miles or 96 km^2) in the Ionian Sea west of continental Greece.

FIGURE 12.3 The great monolith of Uluru (Ayer's Rock) in central Australia: a) aerial view; b) ground view.

Photos by author.

In fiction, the islands are many, but one stands out with events that reflect fate and human frailty. The Aeolian Islands in the Tyrrhenian Sea north of Sicily are the home of Aiolos, the keeper of the world's winds. In sympathy for Odysseus and his plight, Aiolos gives him a leather bag with all but the west wind tightly bound up. As a result, the ship's crew arrive within sight of Ithaca after nine days and nights of smooth sailing. The exhausted Odysseus has held onto the bag the entire way but falls into a deep slumber on final approach. Thinking the bag contains treasure that should be shared out, the crew opens the bag. On release of the winds, the ship is forced back to the Aeolian Islands in a fearsome gale. Aiolos refuses to renew his assistance, dismissing the Greek king as fatefully out of favor with the gods. Odysseus passes down to us through three thousand years of storytelling as a resolute island wanderer but also a character never at a loss for the next plan to reach home.

Another example from mythology concerns the islands off the coast of Norway in the homeland of Petter Dass. Alsten Island is where the seven sisters are frozen in granite. As told in folklore, there were eight troll sisters in all. Well south of Alsten Island, Leka Island represents the eighth and oldest sister. The sisters came out to bathe in a river at dusk and were startled by the troll Hestman (Horse Man), represented by another island at the Arctic Circle (fig. 12.4). The Hestman chases south after the frightened sisters, taking a particular fancy to Leka. Realizing he will not overtake her, the Hestman shoots an arrow to bring her down. Hearing the racket raised by the galloping horse, the Troll King tosses his hat into the air to intercept the arrow and save Leka from injury. The island of Torghatten, with its tunnel cave passing clear through from one side

FIGURE 12.4 Small island (foreground) with monument marking the Arctic Circle off the west coast of Norway. The island in the background is Hestmannøya (Horse Man Island) of Norwegian folk lore.

Photo by author.

to the other, is said to be the result. With the rising sun at dawn, all the principal players in the tale of trolls turn to stone, with Leka halted in her tracks farthest south. In 2010, Leka Island was declared Norway's Geological National Monument as a part of the Trollfjell Geopark. Geologically, the island is highly unusual for igneous rocks that characterize the boundary between the earth's lower crust (gabbro) and the outer mantle (harzburgite). It is extremely rare to encounter the latter ultramafic rock (dominated by the minerals olivine and pyroxene) anywhere on the surface. Moreover, Leka Island is regarded as a small piece of Greenland from the ancestral North American continent welded to Scandinavia during the closure of the Proto-Atlantic Ocean (see chapter 3 on Devonian global geography) but left behind after the subsequent opening of the Norwegian Sea.

ISLANDS WITH NATIONAL PARK STATUS

Several first-class national parks celebrate islands worldwide, attracting visitors for recreation and educational exposure. Four in the U.S. National Park system are well-equipped with campgrounds, good hiking trails, and a range of available literature on the historical and geological aspects of those parks. Acadia National Park on Mount Desert Island in Maine provides a superior setting to experience the intertidal zonation of granite rocky shores under a temperate climate. This part of the Eastern Seaboard is treated by Rachel Carson in *The Edge of the Sea*. The park's natural setting continues to be celebrated by later literary contributions.[11] Sand Beach is a popular bathing spot within the park (fig. 12.5) but is also a locality of notable scientific interest due to the unusually high concentration of carbonate sand derived from the crushed shells and spines of invertebrates like the blue mussel (*Mytilus edulis*) and green sea urchin (*Strongylocentrotus droebachiensis*). The material accumulates faster than can be dissolved by cold, CO_2-enriched seawater.

The Virgin Islands National Park occupies much of St. John and a small part of St. Thomas in the U.S. Virgin Islands, situated east of Porto Rico on the edge of the Caribbean Sea. It offers a stunning contrast in geology and biology to Acadia National Park, with the opportunity for snorkeling and SCUBA diving on coral reefs. The core provenance of limestone formation occurs in warm tropical water between the latitudes of 30°N and 30°S, and the white beaches of St. John at a latitude of 18°20'N are composed almost entirely of coral-derived material. The territory overlaps with the description by Rachel Carson for the

FIGURE 12.5 Sand Beach (foreground) on Mt. Desert Island, Acadia National Park, Maine with granite rocky shores in the background.

Photo by author.

Florida Keys in *The Edge of the Sea*. On the Western Seaboard of the United States, the Channel Islands National Park incorporates five islands offshore the metropolis of Los Angeles in Southern California. Small in size, the islands tell an outsized story related by distinct marine terraces that bear a record of global sea-level change correlated with Pleistocene marine biotas.[12]

Hawaii Volcanoes National Park occupies the heart of the Big Island of Hawaii, featuring the shield volcanoes of Mauna Loa and Kilauea. Mauna Loa ranks as the highest island volcano in the world at 13,678 feet (4,074 m) above sea level. The park also serves as the headquarters for the Hawaiian Volcano Observatory, operated on the rim of the Kilauea Caldera by the U.S. Geological Survey. Spain's Teide National Park on Tenerife in the Canary Islands complements the Hawaiian park in many respects. Regarded as the third-highest island volcano in the world, the great stratovolcano at the center of the Spanish park rises 12,198 feet (3,718 m) above sea level. Its last eruption was in November 1909. Alexander von Humboldt climbed Pico del Teide in 1799 at the outset of his five-year exploration of Spanish colonies in the New World. He left behind a vivid account of his excursion from the town of La Orotava on the island's north coast.[13] Charles Darwin carried a copy of Humboldt's narratives with him on the HMS *Beagle* and was enormously disappointed when plans to stop in Tenerife on the outbound voyage failed to materialize (see chapter 11).

Australia's national government also supports a system of well-organized nature parks throughout the island continent. Windjana Gorge National Park and Geikie Gorge National Park in the Kimberley district of Western Australia are cited in relation to the great Devonian barrier reef (see chapter 5). Uluru-Kata Tjuta National Park celebrates the iconic monolith in the Northern Territory (fig. 12.3).

Ecuador's Galápagos National Park was established in 1959 during the centenary celebration of Darwin's publication *On the Origin of Species*. The park system includes over a dozen islands with a cumulative area of 3,086 square miles (7,995 km²). Only 3 percent of that area is held as private property. Among national parks, the Galápagos has become one of the world's major tourist destinations. Based on data between 2010 and 2015,[14] the number of commercial flights that serviced the islands rose to more than 5,500 from 3,800, and the number of tourist ships cruising the islands increased to seventy-four (with 1,740 berths) from only forty (with 597 berths). The Galápagos park is among the most strictly regulated in the world. Although the number of visitors now exceeds 225,000 per year, the damage is limited by park regulations that mandate the accompaniment of certified park naturalists at a specified ratio to visitors.

MARINE PARKS THAT SHELTER ISLANDS

The Dadohae Haesang Maritime National Park includes the quartzite island of Hongdo or Red Island (see chapter 1) among some additional three thousand islands off the southern coast of the Korean peninsula. Designated as a protected zone in 1981, it is the largest marine park of its kind in South Korea. Ferryboat excursions from the port town of Mokpo are popular with Korean tourists, but guests from elsewhere are still uncommon. Mexico's Loreto Bay National Park in the Gulf of California is a good example of a much smaller marine park that protects several islands within its 80 square miles (206 km²). Tourist activities include deep-sea fishing, scuba diving, kayak tours, and whale watching. Commercial fishing is banned within the park boundaries. Hiking on the islands is largely limited to island shores, where a rich geological and paleontological history may be observed.[15]

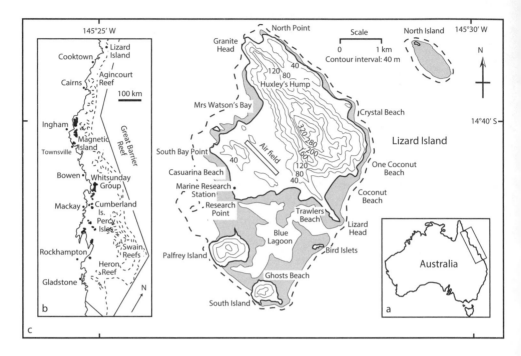

FIGURE 12.6 Maps over the Great Barrier Reef Marine Park off the coast of Queensland: a) Australia with box demarcating the location of the Great Barrier Reef; b) expanded map showing the main towns along the Queensland shore between Cooktown and Gladstone, as well as major islands of the Great Barrier Reef; c) topographic map for Lizard Island and its associated islets in the far north of the Great Barrier Reef.

The largest such entity in the world is undoubtedly the Great Barrier Reef Marine Park (fig. 12.6a and b), which stretches for more than 745 miles (1,200 km) along the shores of Queensland in northeastern Australia. Numerous islands are protected within this area, mainly between Cooktown in the north and Gladstone in the south (fig. 12.6b). Major research stations have headquarters on Lizard Island in the north and Heron Island in the south.

RESORT ISLANDS

Exotic resort islands attract guests to places as distant and varied as found in the Mediterranean (Spain's Ibiza and Mallorca, Italy's Capri, Santorini in Greece), the Caribbean (Antigua, Bermuda, Bahamas, Barbados, Jamaica, Belize, Turks and Caicos), the South Pacific (French Polynesia, Fiji), and the Indian Ocean (Bali, Phuket, Maldives, Seychelles). The seclusion of beautiful beaches, sport activities that include boating and diving, and the promise of fine dining attract vacationers during high seasons that may stretch through many months. The resort business is a huge part of the tourist industry that commands enormous revenues. Aside from the ability of a visitor to afford the privilege, exclusivity is very much a factor based on how remote and pristine any given resort property may be. Two examples that contrast one another are drawn from Australia's Great Barrier Reef Marine Park. In the far north, Lizard Island (fig. 12.6b, c) ranks among the world's most exclusive spots for an elite holiday resort attracting international guests. To the south (fig. 12.6b), Magnetic Island is home to multiple hotels and resorts popular with residents from the coastal towns of Queensland. Both are so-called high islands formed by granite that rises above the surrounding Coral Sea.

Lizard Island (fig. 12.6c) was named by Captain James Cook, who stopped there in August 1770 and climbed to the island summit at 1,065 feet (325 m) above sea level to chart a safe passage for the HMS *Endeavour* through the maze of coral reefs. Today's guests arrive via private aircraft, the nearest commercial airport being at Cooktown, more than 65 miles (~100 km) away (fig. 12b). Offering a limited number of beach-front suites, the Lizard Island resort occupies 2,500 acres (1,013 hectares) around South Bay Point. Sharing the island with many wildlife species, including the large sand monitor (*Varanus gouldii*), resort staff take care to control food waste that might attract unwanted feeding. All food scraps from the kitchen operation are frozen and shipped to the mainland. Resort guests rarely venture to the east side of the island, where biologists from the Marine Research Station can enjoy equally spectacular beaches. The aboriginal owners of the island hail from the Dingaal group, who still regard the island as a sacred place. According to oral tradition, the earliest elders from perhaps fifty thousand years ago could walk to the island from what is now the mainland. If true, it means that Lizard Island was first a mainland monadnock before it became a drowned monadnock.

Magnetic Island is a much larger island (20 square miles or 52 km²), taking its name from magnetic anomalies that apparently affected the compass aboard the HMS *Endeavour* as it sailed by in 1770. The island is serviced by ferryboats from Townsville, less than 5 miles (~8 km) away. Around two thousand permanent residents live on the island, but the population easily doubles during the high season. The island itself is pleasant enough, but water clarity around it is poor due to the high sediment load carried by the Ross River that empties into the sound in front of the city. Sugarcane plantations have replaced the native rainforest upstream from Townsville, and substantial outwash of red soil occurs during the wet season. The dark plume of sediment-laden water flowing into the ocean is readily apparent from the air. In contrast, Lizard Island is 19 miles (~30 km) off the nearest mainland coast, and water clarity does not suffer at that distance.

PRISON ISLANDS

Remoteness from mainland populations also affects the location of island prisons. Quite the opposite of resort islands that enjoy seclusion, prison islands are ranked by their notoriety for harsh punishment. A functioning prison island in the Mexican federal penal system is located on Maria Madre Island, housing approximately 2,500 prisoners. The island is about 60 miles (98 km) off the Mexican mainland between Mazatlán and Puerto Vallarta. Several deactivated prisons have historic reputations. In the U.S. penal system, Alcatraz Federal Penitentiary was a maximum, high-security prison on Alcatraz Island in San Francisco Bay, only 1.25 miles (2 km) offshore the city of San Francisco. After twenty-nine years of operation, the prison was closed in 1963 but now functions as a museum. Cold water and strong tidal currents were major deterrents against escape. Many escape attempts were made, but only five prisoners made it out, never to be heard from again. No bodies were recovered, and the escapees are thought to have drowned.

Tarrafal Prison on Santiago Island in the Cape Verdes was the much-feared home to political dissidents during the regime of Portugal's António Oliveira Salazar during his thirty-six years of power, ending in 1968. The site is now a museum operated by the Republic of Cape Verde, which received independence from Portugal in 1975. Now also a museum, South Africa's Robben Island is where Nelson Mandela was imprisoned for eighteen years before he became the country's first president after the fall of the apartheid state. The name of the island comes from the Dutch for "Seal Island," and it is located 4.5 miles (~7 km) off Cape Town on the continental mainland. The most infamous of all island prisons was Devil's Island, as it was called by its inmates. Founded in 1852 and closed only in 1953, the official name for the penal colony was Bagne de Cayenne, after the capital of French Guiana in South America. The main prison covers 35 acres (14 hectares) and is 9.25 miles (15 km) off the continental mainland. An estimated eighty thousand Frenchmen were sent there to live under harsh conditions and a high death rate.

An Ecuadoran penal colony was established in 1944 near Puerto Vilamil on Isabela Island in the Galápagos Islands. Closed after thirteen years, the initial prison population amounted to three hundred inmates supervised by thirty guards. With little else to keep the prisoners occupied, wardens found a way to keep them busy with the construction of a wall that stretches some 328 feet (100 m) across a narrow valley. The wall is all that remains of the prison camp (fig. 12.7). The structure for which the only purpose was to impose hard labor is called the Wall of Tears.

FIGURE 12.7 The Wall of Tears at the former penal colony on Isla Isabela in the Galápagos Islands of Ecuador.

Photo by author.

ISLANDS AS MILITARY TESTING GROUNDS

Remote and sparsely populated islands have a dark history of use by military planners as testing grounds for explosive ordinance. The United States Navy began to use the Island of Vieques off Puerto Rico as an ammunition depot and testing range in 1941. Approximately two-thirds of the island (22,000 acres or 8,900 hectares) was subjected to ship-to-shore bombardment, air-to-ground bombing, and live-fire training for amphibious landings. Operations ceased in 2003 after many years of protests by Puerto Ricans. A civilian population continued to live on that part of the island not occupied by the U.S. Navy, and medical tests have reported a high level of garden contamination by heavy metals.

Eniwetok Atoll in the Marshall Islands (see fig. 11.2) consists of forty individual islands around a central lagoon with a diameter of 20 miles (32 km). Operation Ivy was carried out on November 1, 1952, when the U.S. military detonated a hydrogen bomb 30,000 feet (9,144 m) over the atoll. A second test was performed soon thereafter, with a nuclear device exploding from the top of a tower 200 feet (61 m) above one of the atoll's islands. More than forty additional blasts were conducted, ending in 1958. Although the atoll's population was evacuated beforehand, nuclear explosions on Eniwetok resulted in an increased incidence of thyroid cancer among Marshallese people living elsewhere. The United States and the Marshall Islands government reached an agreement in 1983, setting up a compensation fund. As late as 2000, soil tests in the Marshall Islands continued to show levels of contamination by radioactive fallout.

Preparations for the Eniwetok bomb tests by the U.S. military required a geological survey based on a drilling program that resulted in a significant scientific breakthrough regarding Darwin's hypothesis of atoll development. After cutting through 4,200 feet (1,280 m) of coral limestone, the drill bit encountered basalt bedrock.[16] Microfossils retrieved from the overlying limestone showed conclusively that reef growth commenced more than thirty million years ago around a sinking island during the Eocene epoch. Upward reef growth continued at an average rate of an inch (2.5 cm) per one thousand years, long after the volcano disappeared on that part of the subsiding Pacific Tectonic Plate.

ISLAND NATURE SANCTUARIES

In contrast to sending humans to prison islands with little chance of escape, nature sanctuaries protect islands where humans are kept out so that nature may operate under its own terms without external interference. The system with which I am most familiar operates under the jurisdiction of the Mexican federal government as island biosphere reserves protecting forty or so named islands in the Gulf of California under a decree issued in 1978. With a composite island territory of 1,150 square miles (2,977 km²), the reserves were reinforced by language in 2000 formally declaring the islands as "Areas of Protection of the Flora and Fauna."[17] Regulations under which scientific research is allowed come under scrutiny by government officials who issue the necessary permits.

The Hawaiian Islands National Wildlife Refuge and the Midway Atoll National Wildlife Refuge were declared a Marine National Monument in June 2006 and enlarged in 2016 to cover 582,000 square miles (1,510,000 km²). Islands and atolls within the zone fall under regulations mandated by various federal and state agencies. The monument's oceanic boundary extends outward from islands for 200 miles (320 km) under regulations by the U.S. National Oceanic and Atmospheric Administration. Another large sanctuary system is the Pacific Remote Islands Marine National Monument, which was constituted in January 2009 and expanded in 2014 to embrace 490,340 square miles (nearly 1,270,000 km²) in the south-central Pacific Ocean. Tropical areas protected from commercial fishing, waste dumping, and deep-sea mining include marine zones around Baker Island, Howland Island, Jarvis Island, Wake Island, Johnston Atoll, Palmyra Atoll, and Kingman Reef. In combination, these two systems under American protection represent the planet's most extensive nature refuge.

In Russia, similar nature reserves with strictly controlled access for scientific research fall under the classification of a *zapovednik*. One such reserve covers all of Wrangel Island (2,900 square miles or 7,600 km²) in the Siberian Arctic Ocean. The island is renowned as the last holdout of woolly mammoths (*Mammuthus primigenius*), believed to have survived until about 1650 BCE.

GEOHERITAGE, GEOPARKS, AND PALAEOPARKS

Geoheritage suggests that certain places possess geologic attributes with an extraordinary shared value for humans everywhere. Specific sites that embody distinctive rock or mineral formations or unusual fossils fall into a category that merits protection for their scientific and educational value. Places where geological features played an important role in cultural or historical events represent another. Where the landscape is especially attractive due to its geological underpinnings, geoheritage constitutes yet a third category. The worthiness of geopark status is especially compelling where all three aspects appear in the same setting. Under the United Nations Educational, Scientific and Cultural Organization (UNESCO), a procedural system exists for the proposal and approval of geoparks united with common attributes and management goals in a global network. The UNESCO Global Geoparks program was launched in 2000 and attempts to emulate the success already achieved by national parks and monuments in countries such as the United States and Australia. Essential to the creation of a geopark are requirements that the proposed area exhibits some combination of geoheritage with the potential to generate economic activities in local communities through sustainable ecotourism. Island geoparks sanctioned by UNESCO are up and running in Greece (two parks), Cyprus, Croatia (Vis Archipelago), Sicily, Sardinia, the Canary Islands (El Hierro Island), the Azores, Iceland (two parks), Norway (a national geological monument), South Korea (Jeju Island), Japan (Oki Islands), and Indonesia (two parks).

Among the monadnocks and paleoislands described in this volume, protection under government aegis is already strong, and there is future potential for the development of geoparks. Mount Monadnock in New Hampshire and the Baraboo monadnocks of Wisconsin are entirely or partially within state parks, as are parts of the Black Hills in South Dakota. South Korea's Hongdo (Red Island) is part of a maritime national park. Mount Misen on Japan's Miyajima is a protected parkland. Prospects exist for the formal development of geoparks related to the Ordovician Jens Munk Archipelago in Churchill, Canada (chapter 3), the Devonian Mowanbini Archipelago in the Oscar Range of Western Australia (chapter 5), the Cretaceous Eréndira Islands of Mexico's Baja California (chapter 8), Miocene Porto Santo of Madeira (chapter 9), as well as the Pleistocene shores of Santiago visited by Charles Darwin in the Cape Verde Islands (chapter 11). Santa Maria Island in the Azores now enjoys a high level of protection as the world's first palaeopark (chapter 10).

MEANING ASCRIBED TO PALEOISLANDS

Paleoislands are fossil islands that reflect physical and biological conditions pertaining to ecosystems that functioned during the geologic past as microcosms largely isolated from the greater worlds around them. They are time capsules that allow the visitor to experience conditions as they were at different stages in the evolution of Planet Earth. Like many islands today, well-preserved paleoislands tend to be small. This makes them readily accessible to the visitor both physically and intellectually. The time slices explored in this volume (see fig. 1.9) represent a minuscule sample of our planet's immense history. Former shorelines, particularly those that may be traced around ancient islands, are not such a rare curiosity as might be expected. Anyone who can learn to read a geologic map and understand the pattern of older rocks encircled by younger marine strata may set out to locate and explore paleoislands on their own account.

Paleoislands alone have no meaning. But they exist for the human species (*Homo sapiens*) to discover and contemplate. As implied by Rachel Carson, monadnock heights and related paleoislands are not reserved for the enjoyment of geologists alone. They are free for all to experience either in person or vicariously through the narratives of others, including the likes of Charles Darwin and James D. Dana (chapter 11). The nineteenth-century world they inhabited was only beginning to suffer the ill effects of pollution from an industrial age with an insatiable appetite for fossil fuels. Our world has advanced precariously down that dark road to the extent that global warming has become a serious danger.[18] Among the findings published through the U.S. Global Change Research Program in its Fourth National Climate Assessment is the assessment that: "Rising water temperatures, ocean acidification, retreating arctic sea ice, sea level rise, high-tide flooding, coastal erosion, higher storm surge, and heavier precipitation events threaten our oceans and coasts."[19] Public response to the threat of climate change has been slow to build, but the United Nations Office for Disaster Risk Reduction reports that 75 percent of the U.S. population acknowledges that such change is happening based on exposure to a record number of heatwaves, wildfires, and hurricanes.[20] Even so, it is projected that the severity of such events will continue to worsen over the coming years.

Limited as the geological record is concerning paleoislands, it is clear enough that global warming occurred during intervals in the distant past long before the arrival of humans. Paleoislands as distant from one another as the Azores in the North Atlantic and Mexico's Gulf of California (chapter 10) were subject to unusually intense storms during the Pliocene warm period. The Baraboo Archipelago of Wisconsin was indisputably lashed by hurricanes much earlier during the Cambrian period (chapter 2). The difference is that nature modulated itself over intervals of tens of thousands of years or longer, whereas human activities during only the last decades have quantifiably increased the volume of heat-trapping gases in Earth's atmosphere more rapidly than at any other time in the planet's history. Humans have become a major geological force during the relatively short time some have called the Anthropocene.[21] The ultimate value and meaning of islands caught in deep time are that they provide an opportunity for us to appreciate the antiquity of Planet Earth, so deserving of our veneration and resolution to make ourselves better stewards of its health.

FINAL SUMMARY

Islands today have come to mean many different things to our kind. They have been abused as places where fearful weapons capable of obliterating society were tested during the Cold War. Not so much at present, they have been used as places to banish criminals and political dissidents. At the opposite extreme, certain islands and surrounding marine reserves are set aside as places where human access is prohibited so that nature may heal itself. Notably, some islands, and not a few paleoislands, have a reputation as sacred places where humans may find a spiritual connection with nature. Like the Kobo Daishi, Ralph Waldo Emerson, and countless others such as Nicholas Black Elk, let us join hands to descend Mount Misen together. Let us return, each to our own homes like a penitent Odysseus, with a renewed determination to seek better ways for the advancement of our society in more perfect harmony with nature.

GLOSSARY

ACADIAN OROGENY. The name comes from French settlements in the northeastern parts of North America, but geologically refers to the closing of the proto-Atlantic Ocean in Devonian time, when the ancestral North American and European continents collided and formed a major mountain chain across a united Euramerican paleocontinent.

ANDESITE. Igneous rocks from surface flows (volcanic lavas and breccias) rich in minerals plagioclase and lesser amounts of pyroxene, hornblende, and/or biotite. Although a freshly broken surface is generally dark gray in color, weathered surfaces often take on a reddish cast due to oxidation of iron content. The name is derived from the Andes, which is representative of the mountain belt where these rocks occur due to ocean-plate subduction against a continental margin.

ANTICLINE. A structural deformation of the Earth's crust in which rock layers are compressed to make a positive fold with sides that slope downward away from a fold axis.

ATOLL. A circular line of low-lying islands with a central lagoon that formed originally around a volcano subject subsidence of a tectonic plate. The islands are enclosed by a barrier reef that grew upward as volcanic basement rocks sank lower in the ocean.

BARNACLE. Marine invertebrate, classified within the phylum Arthropoda, solitary in habit, attached to rocks via a cemented basal plate surrounded by flexible plates configured in a dome shape. A feeding appendage (cirri) is extruded into the water column to sweep up food particles in suspension.

BASALT. Igneous rocks of an extrusive origin (flood lavas or submarine pillow lavas) rich in minerals plagioclase, pyroxene and often olivine. These rocks typically form on the ocean floor or in continental rifts. They are dark and generally weather dark.

BASEMENT ROCKS. Metamorphic and/or intrusive igneous rocks overlain at depth by sedimentary rocks.

BEDROCK. Solid rock, stratified or not, that occurs beneath soil, gravel, or any other unconsolidated surface materials.

BRACHIOPOD. Shelled marine invertebrate with two valves ($CaCO_3$) that are typically different from one another in shape, but individually bilaterally symmetrical. Phylum Brachiopoda.

BRECCIA. A kind of conglomerate in which the individual clasts are rough and angular.

CAMBRIAN. A geological period, roughly 54 million years in duration, that began about 542 million years ago and terminated close to 488 million years ago. All those sedimentary and igneous rocks that originated during that interval are said to belong to the Cambrian System.

CEPHALOPOD. A group of marine invertebrates classified among the mollusks that includes squids and octopi as well as shelled animals like the pearly nautilus and similar antecedents from the fossil record.

CLAST. An individual fragment of rock (varying sizes), eroded by the action of wind, waves, or running water from a parent source.

CONGLOMERATE. A sedimentary rock composed of cemented cobbles and boulders eroded from pre-existing rocks which may be different from one another depending on how far away the source was that supplied the clasts.

CONODONT. Extinct chordate animal with a bilaterally symmetrical body lacking vertebrae but with phosphatic, cone-shaped "teeth" in the mouth area used to grasp small prey items prior to ingestion. Rapid evolution of the animal's hard parts as microfossils makes them useful for correlation of Paleozoic strata (Cambrian to Permian systems).

COQUINA. Bedded accumulations of cemented shells that typically exclude intermittent sediment.

CORAL. Marine invertebrates that are solitary or colonial and belong to the phylum Coelenterata. Many species secrete a solid skeleton ($CaCO_3$) and are the principal contributors to the construction of reefs.

CORALLINE RED ALGAE. Marine plants that belong to the division Rhodophyta and have the ability to secrete skeletons of calcium carbonate. The adjective "coralline" is applied to indicate that the algae mimic corals in appearance.

CRETACEOUS. A geological period, roughly 80 million years in duration, that began 145.5 million years ago and terminated 65.5 million years ago. All those sedimentary and igneous rocks that originated during that interval are said to belong to the Cretaceous system.

DENDRITIC DRAINAGE. Streams that merge together in a branch-like pattern at acute angles.

DEVONIAN. A geological period, roughly 57 million years in duration, that began about 416 million years ago and terminated close to 359 million years ago. All those sedimentary and igneous rocks that originated during that interval are said to belong to the Devonian system.

DOLOSTONE. A sedimentary rock dominated by bioclastic grains of $CaCO_3$ but also infused with magnesium (Mg) typically from an inorganic source. A pure limestone may be changed secondarily to a dolostone by the extensive substitution of magnesium for calcium.

DROP STONES. Ice rafted stones carried to sea from continents by way of glacial ice. When the sea ice melts, the stones drop to the bottom of the ocean.

ECHINOID. Marine invertebrate, also called a sea urchin, that is solitary in plan, exhibits five-fold symmetry, and belongs to the phylum Echinodermata. All species secrete a shell (test) formed by calcite ($CaCO_3$).

ECOSYSTEM. A physical setting such as a lake, delta, or shallow sea together with all the inter-related organisms that thrive under the specific limitations of that setting.

EUSTASY. The concept that sea level may rise or fall on a global basis simultaneously all around the world due to additions or subtractions of ocean water related to major glaciations, but also due to other geophysical factors that change the configuration of ocean basins.

FACIES. Sedimentary rocks and fossils that represent contemporaneous variations in a lateral continuum. A common representation relates to facies changes in an onshore-offshore pattern.

FAUNA. The animals that live in a given area or environment. A faunal list gives the names of those animals (or fossils) found in a given habitat.

FORESET BEDS. Sediments deposited in inclined layers by rivers on delta lobes on reaching the ocean.

FREEBOARD. A nautical term for the distance between a ship's waterline and its upper deck, but also applicable to continents which float at different levels in the earth's crust.

GEOHERITAGE. Features of geology, small and large including entire landscapes, that are intrinsically valuable to science but with cultural ramifications providing insights into the history of planet Earth.

GEOMORPHOLOGY. The study of physical landforms and the natural processes that lead to their development at the surface of the Earth.

GNEISS. A metamorphic rock with bands of coarse grains that may or may not be folded by compression.

GRANITE. Igneous rock that cooled far underground with large mineral crystals that typically include feldspar, plagioclase, quartz, and biotite.

HADLEY CELLS. Atmospheric circulation marked by upward convection of warm, wet air around the equator and downward convection of cool, dry air typically around 30° N and S of the equator in regions where the world's deserts are located.

IGNEOUS ROCKS. Rocks cooled from molten material either deep within the Earth's crust (intrusive) or at the Earth's surface (extrusive) as a result of magmatic activity.

INLIER. A body of older rocks reduced in size by erosion and subsequently encircled by younger sedimentary rocks that form an unconformity against the pre-existing rocks.

INTERTROPICAL CONVERGENCE ZONE. The geographic zone near the equator where trade winds converge typically from the northeast and southeast.

ISLAND BIOGEOGRAPHY. The study of islands of all sizes with respect to their colonization by species from an adjacent mainland that may be close by or much farther away.

JOINT. A vertical fracture in rocks.

JURASSIC. A geological period, roughly 54 million years in duration, that began about 199.5 million years ago and terminated 145.5 million years ago. All those sedimentary and igneous rocks that originated during that interval are said to belong to the Jurassic system.

KARST. A range of landforms that develop both on and within terrain dominated by limestone cover due to dissolution of $CaCO_3$ under a humid climate. The name comes from the Karst district on the coast of the Adriatic Sea.

LIMESTONE. Sedimentary rock consisting of calcium carbonate ($CaCO_3$) derived primarily from organic remains of marine invertebrates such as corals, mollusks, echinoderms, and coralline algae.

MARINE TERRACE. A narrow coastal rim that usually slopes gently seaward and is veneered by a marine deposit. Formation of the terrace is caused by intertidal erosion and the position of the terrace depends on changes in global sea level with respect to changes in the local or regional elevation of the coastline.

METAMORPHIC ROCKS. Rocks either sedimentary or igneous in origin that are subsequently altered by heat and pressure due to deep burial in the Earth's crust. Limestone may be altered to marble, for example, and granite may be altered to schist.

MIOCENE. A geological epoch, roughly 18 million years in duration, that began about 23 million years ago and terminated a little more than 5 million years ago. All those sedimentary and igneous rocks that originated during that interval are said to belong to the Miocene Series.

MOLLUSK. Marine invertebrates that are solitary in plan and belong to the phylum Mollusca. The phylum includes chitons (class Polyplacophora) land and sea snails (class Gastropoda), clams (class Bivalvia), as well as squids and the octopus (class Cephalopoda). In particular, the shelled gastropods and bivalves lend themselves to fossilization.

MONADNOCK. A hill or mountain composed of extremely hard rock that resists erosion and stands above the surrounding landscape where the underlying bedrock is more susceptible to erosion.

ORDOVICIAN. A geological period, roughly 45 million years in duration, that began about 488 million years ago and terminated close to 444 million years ago. All those sedimentary and igneous rocks that originated during that interval are said to belong to the Ordovician system.

OUTLIER. Isolated bodies of stratified rock detached from the main outcrop due to erosion of the surrounding area between the outlier and the rest of the outcrop. Outliers typically form buttes or mesas that may be far removed from similar rocks.

PALEOECOLOGY. Study of ecology represented by ancient ecosystems.

PALEOISLAND. An ancient island, the origin of which dates back thousands to millions of years ago and may no longer be surrounded by water.

PERMIAN. A geological period, roughly 58 million years in duration, that began about 299 million years ago and terminated close to 251 million years ago. All those sedimentary and igneous rocks that originated during that interval are said to belong to the Permian System.

PHYLLITE. A metamorphic rock with high clay content altered under heat and pressure from mudstone, but also close to schist as a metamorphic grade.

PILLOW BASALT. Basalt magma extruded under water that forms "pillow-shaped" bubbles that harden as solid rock on cooling.

PLEISTOCENE. A short geological epoch, dating from about 2,588,000 years and ending about 10,000 years ago, that bridges the prior Pliocene epoch and the Holocene (Recent). All sedimentary and igneous rocks that originated during that time interval are said to belong to the Pleistocene Series.

PLIOCENE. A geological epoch, roughly 2.8 million years in duration, that began more than 5 million years ago and terminated about 2,588,000 years ago. All those sedimentary and igneous rocks that originated during that interval are said to belong to the Pliocene Series.

QUARTZITE. A metamorphic rock that results from the conversion of sandstone under high heat and pressure such that the original sand grains are fused together.

RHODOLITH. A particular kind of coralline red algae that grows unattached on the seafloor. The rhodolith assumes a spherical shape due to frequent movement with wave and current activity during the lifetime of the alga. The alga may colonize a tiny piece of shell or a rock fragment as large as a pebble, thereafter growing outward in a radial pattern.

RHYOLITE. A volcanic rock formed as a surface flow that is chemically the fine-grained equivalent of granite.

RIFT VALLEY. A region, typically linear in demarcation, where a continent has begun to break apart or where ocean crust continues to spread apart in opposite directions.

ROPEY LAVA. Lava characterized by a wrinkled texture that results as the outer surface cools or stiffens while more fluid magma below continues to flow. Also known by the Hawaiian term pahoehoe lava.

RUDIST CLAM. An extinct bivalve belonging to the phylum Mollusca commonly distinguished from other clams by their extreme asymmetry with one valve enlarged in the form of a cone or large cup and the opposing valve forming a lid.

SCREE. Pile of rock waste found at the base of a cliff or a sheet of coarse debris that covers a cliff or steep mountainside.

SEDIMENTARY ROCKS. Rocks formed by the burial and cementation of inorganic sediments like pebbles, sand, silt, and clay, or the fragments of broken corals and shells that form.

SHIELD VOLCANO. Large, shield-shaped volcano that builds up radially from low-viscosity lava flows. The Big Island of Hawaii is the ideal example for this class of volcanos.

SILURIAN. A geological period, roughly 28 million years in duration, that began about 444 million years ago and terminated 416 million years ago. All those sedimentary and igneous rocks that originated during that interval are said to belong to the Silurian system.

SKY ISLAND. Generic term for a monadnock representing a resistant body of rock that rises high above the surrounding countryside and provides living space for plants and animals different from those on the plains below.

STRATA. Layered sedimentary rocks.

STRATOVOLCANO. Steep-sided volcanos often high in elevation that add layers of ash and other ejecta to the sides in a cone-shaped profile. The Italian island of Stromboli is the ideal example for this class of volcanos.

STROMATOLITE. A microbial deposit made by cyanobacteria, typically laminar in organization, originating far back in Precambrian time. The name derives from the Latin *stroma* for bed; and *lithos* for stone.

STROMATOPOROID. A kind of sponge with calcified layers penetrated by pores for the intake and expulsion of sea water. Same Latin root (*stroma*) as applied to stromatolites.

SURTSEYAN VOLCANO. A volcano that reaches to the surface from beneath the sea with violent eruptions that typically send large volumes of volcanic ash into the air. The Icelandic island of Surtsey is the ideal example for this class of volcanos.

SYNCLINE. A structural deformation of the Earth's crust in which rock layers are compressed to make a negative fold with sides that slope upward away from a central fold axis.

TACONIC OROGENY. Part of the Appalachian Mountain chain, the worn remnants of the Taconics in New England lend their name to an episode of mountain building during the late Ordovician and early Silurian periods when the proto-Atlantic Ocean slowly disappeared due to subduction of ocean crust beneath the ancestral North American continent.

TECTONIC PLATES. Dynamic plates at the surface of the Earth defined by boundaries where the crust is pulling apart at oceanic spreading zones or continental rift zones, as opposed to places where dense oceanic crust is subducted beneath less dense continental crust along oceanic trenches.

TETHYS SEAWAY. A former east-west seaway that extended fully around the world during the Cretaceous period, now represented only locally by the Mediterranean Sea.

TRACE FOSSIL. Signs of life left behind in sediments and sedimentary rocks as tracks, trails, burrows, or borings.

TRELLIS DRAINAGE. Streams that merge together at right angles, typically due to structural fold patterns in the underlying rocks.

TRILOBITE. An extinct arthropod distinguished by a three-part organization with a head shield, flexible middle section (thorax), and tail shield.

TUFF. A rock formed from volcanic ash and small fragments (usually less than 4 mm or 1/8 in diameter) of volcanic rock blasted by an eruption.

UNCONFORMITY. A surface of erosion that separates two bodies of rock and represents an interval of time during which deposition ceased, some material was removed, and then deposition resumed again. An *angular unconformity* involves a set of tilted rock layers below the unconformity surface overlain by mainly flat-lying strata above. Other types of unconformities may involve a juncture between igneous or metamorphic rocks below the unconformity surface and stratified sedimentary rocks above.

NOTES

1. HOW TO LISTEN TO A SKY ISLAND WITH GLOBAL AMBITION: CLIMBING MOUNT MONADNOCK

1. The geology of Mount Monadnock is treated in technical detail by the structural geologist Peter J. Thompson (University of New Hampshire) in the *Guidebook for Geological Field Trips in New England: 2001 Annual Meeting of the Geological Society of America, Boston, Massachusetts*, ed. D. P. West Jr. and R. H. Bailey (Boston: Geological Society of America, 2001), R1–R17. The type location for the Littleton Formation occurs in Maine, where fossils typical of Lower Devonian strata are present. Because of metamorphic deformation, the age-equivalent rocks around the summit of Mount Monadnock were altered from mainly sandstone to quartzite.

2. The standard mineralogy reference book is *Dana's Manual of Mineralogy*, first issued in 1848 by the Yale professor, James Dwight Dana (1813–1895) and used by generations of geology students in subsequent editions down to the present day. Sillimanite is a rare silicate mineral after aluminum and aluminum-oxide found in high-grade metamorphic rocks like schist. The mineral was named in honor of Benjamin Silliman (1779–1864), under whom Dana studied mineralogy at Yale College.

3. Ralph Waldo Emerson was the principal architect of transcendentalism, a uniquely American philosophical concept loosely related to pantheism. Among other influences, quasireligious aspects of Hinduism took hold of Emerson, who was the son of a Unitarian minister and served as an ordained Unitarian pastor in Boston from 1829 to 1832. Emerson's *Nature* (1836) espoused the belief that "all science has one aim, namely, to find a theory of nature" and encouraged humans to "enjoy an original relation to the universe."

4. A. W. Grabau, *The Rhythm of the Ages* (Peking: Henry Vetch, 1940), 561.

5. M. E. Johnson and B. G. Baarli, "Geomorphology and Coastal Erosion of a Quartzite Island: Hongdo in the Yellow Sea off the SW Korean Peninsula," *Journal of Geology* 121 (2013): 503–16.

6. Johnson and Baarli, "Geomorphology and Coastal Erosion."

7. See J. R. Ali, "Islands as Biological Substrates: Classification of the Biological Assemblage Components and the Physical Island Types," *Journal of Biogeography* 44, no. 5 (May 2017): 984–94, https://doi.org/10.111/bji.12872.

8. Grabau, *The Rhythm of the Ages*.

9. M. E. Johnson and B. G. Baarli, "Development of Intertidal Biotas Through Phanerozoic Time," in *Earth and Life*, ed. J. A. Talen (Dordrecht, Netherlands: Springer Science + Business Media, 2012), 63–128.

10.   Johnson and Baarli, "Development of Intertidal Biotas."

11.   The classical Greek roots for the term *eustasy* derive literally from the words for "well" and "standing" as defined by the Austrian geologist Edward Suess (1831–1941) in his book *Das Antitze der Erde* (The face of the earth), published serially beginning in 1885.

2. HOW AN ISLAND CLUSTER ACQUIRES ITS SHAPE: A JOURNEY IN LATE CAMBRIAN TIME TO WISCONSIN'S BARABOO ARCHIPELAGO

1.   R. A. Davis, "Precambrian Tidalites from the Baraboo Quartzite District, Wisconsin," *Marine Geology* 235, no. 1 (December 2006): 247–53.

2.   The Wisconsin geologists I. W. D. Dalziel and R. H. Dott Jr. are the undisputed experts on the geology of the Baraboo District based on their study of the "Geology of the Baraboo District " published in the *Geological and Natural History Survey, Information Circular* 14: 1–163 (1970). Refer to plate 6 in their publication to appreciate the full range quartzite clast sizes eroded during Cambrian time.

3.   I am indebted to an uncle, the late David J. Carey, a graduate student at the University of Wisconsin in Madison in the early 1960s, who took me on my first visit to Parfrey's Glen. The locality left a lasting impression and helped spur my interest in geology.

4.   Approved by the IUGS in 2003, the base of the series is defined by a Global Stratotype Section and Point (GSSP) at the first occurrence of a particular trilobite species (*Glyptagnostus reticulatus*) within a thick succession of limestone beds from the Huaqiao formation in the Wuling Mountains of northwestern Hunan Province, China. The word *furong* in Mandarin Chinese translates as "lotus flower," which alludes to Hunan as China's lotus-flower province.

5.   S. Marshak, M. S. Wilkerson, and J. DeFrates, "Structural Geology of the Baraboo District," in *Geology of the Baraboo Wisconsin Area*, ed. R. Davis Jr., R. H. Dott Jr., and I. W. D. Dalziel 13–36: *Geological Society of America Field Guide* (2016), 43, 81.

6.   R. H. Dott Jr., "Cambrian Tropical Storm Waves in Wisconsin," *Geology* 2, no. 5 (May 1974): 243–46.

7.   J. D. Eoff, "Sequence Stratigraphy of the Upper Cambrian (Furongian, Jiangshanian and Sunwaptan) Tunnel City Group, Upper Mississippi Valley: Transgressing Assumptions of Cratonic Flooding," *Sedimentary Geology* 302 (2014): 87–101.

8.   Reassessment of Cambrian wave height necessary to erode the largest quartzite boulders is provided by R. H. Dott Jr. and C. W. Byers, "Cambrian Geology of the Baraboo Hills," *Geological Society of America Field Guide 43* (2016): 47–54.

9.   Dott Jr., "Cambrian Tropical Storm Waves in Wisconsin."

10.   The distinguished career of Dr. Christina Lochman-Balk is memorialized in a tribute under the heading "the women of tech" published in the New Mexico Tech alumni magazine *Gold Pan* (Summer 2014): 21–22.

11.   C. Lochman-Balk, "Upper Cambrian Faunal Patterns on the Craton," *Geological Society of America Bulletin* 81 (1970): 3197–3224.

12.   J. W. Hagadorn, R. H. Dott Jr., and D. Damrow, "Stranded on a Late Cambrian Shoreline: Medusae from Central Wisconsin," *Geology* 30 (2002): 147–50.

13.   M. E. Johnson and B. G. Baarli, "Development of Intertidal Biotas Through Phanerozoic Time," in *Earth and Life*, ed. J. A. Talen (Dordrecht, Netherlands: Springer Science + Business Media, 2012), 63–128.

14.   C. Lochman-Balk, "Paleo-Ecological Studies of the Deadwood Formation (Cambrian-Ordovician)," In *Paleontology and Stratigraphy*, ed. R. K. Sundaram (Proceedings of Section 8 of the 22nd International Geological Congress, India, 1964), 25–38.

15. M. E. Johnson, M. A. Wilson, and J. A. Redden, "Borings in Quartzite Surf Boulders from the Upper Cambrian Basal Deadwood Formation, Black Hills of South Dakota," *Ichnos* 17 (2010): 48–55.

16. R. N. Donovan and M. D. Stephenson, "A New Island in the Southern Oklahoma Archipelago," *Oklahoma Geological Survey Circular* 92 (1991): 118–21.

3. HOW ISLANDS TRADE IN PHYSICAL WEAR AND ORGANIC GROWTH: A JOURNEY IN LATE ORDOVICIAN TIME TO HUDSON BAY'S JENS MUNK ARCHIPELAGO

1. M. L. Droser and S. Finnegan, "The Ordovician Radiation: A Follow-Up to the Cambrian Explosion," *Integrative and Comparative Biology* 43 (2003): 178–84; C. M. Ø. Rasmussen et al., "Cascading Trend of Early Paleozoic Marine Radiations Paused by Late Ordovician Extinctions," *Proceedings of the National Academy of Sciences* 116, no. 15 (2019): 7207–13.

2. M. E. Johnson, D. F. Skinner, and K. G. MacLeod, "Ecological Zonation During the Carbonate Transgression of a Late Ordovician Rocky Shore (North-Eastern Manitoba, Hudson Bay, Canada)," *Paleogeography, Paleoclimatology, Paleoecology* 65 (1988): 93–114.

3. The conodont animal is described as a bilaterally symmetrical, eel-shaped swimmer based on Carboniferous material studied in D. Briggs, E. Clarkson, and R. Aldridge, "The Conodont Animal," *Lethaea* 16 (1983): 1–14.

4. Approved by the International Union of Geological Sciences (IUGS) in 2006, the base of the Hirnantian stage is now defined by a Global Stratotype Section and Point at the first occurrence of a graptolite species (*Normalograptus extraordinarius*) within thick shale beds north of Yichang near the village of Wangjiawan in Hubei Province, China.

5. Details on the correlation of the Hirnantian stage in the central part of North America that includes a distinct carbon isotope marker are summarized by M. W. Demski et al., "Hirnantian Strata Identified in Major Intracratonic Basins of Central North America: Implications for Uppermost Ordovician Stratigraphy," *Canadian Journal of Earth Sciences* 52 (2015): 68–76.

6. Discovery of the first rock-encrusting corals from the Upper Ordovician of Hudson Bay was made by my fellow geologist and spouse, B. Gudveig Baarli. Only a few months before the expedition to Churchill, we had enjoyed a holiday trip to the British Virgin Islands where an indelible impression was made by living corals in aquamarine water encrusted on red granite. The experience swimming in tropical waters among those corals remained very fresh in mind, prompting the fossil find.

7. M. E. Johnson and B. G. Baarli, "Encrusting Corals on a Latest Ordovician to Earliest Silurian Rocky Shore, Southwest Hudson Bay, Manitoba, Canada," *Geology* 15 (1987): 15–17.

8. R. J. Elias and G. A. Young, "Enigmatic Fossil Encrusting an Upper Ordovician Rocky Shore on Hudson Bay, Canada, Is a Coral," *Journal of Paleontology* 74 (2000): 179–80.

9. Johnson and Baarli, "Encrusting Corals."

10. Upper Ordovician medusa (jellyfish) from Manitoba are described in G. A. Young et al., "Exceptionally Preserved Late Ordovician Biotas from Manitoba, Canada," *Geology* 35, no. 10 (2007): 883–86.

11. Johnson, Skinner, and MacLeod, "Ecological Zonation During the Carbonate Transgression."

12. The largest trilobite is described in D. M. Rudkin et al., "The World's Biggest Trilobite—*Isotelus rex* New Species from the Upper Ordovician of Northern Manitoba, Canada," *Journal of Paleontology* 77, no. 1 (2003): 99–112.

13. D. F. Skinner and M.E. Johnson, "Nautiloid Debris Oriented by Long-Shore Currents Along a Late Ordovician–Early Silurian Rocky Shore," *Lethaia* 20 (1987): 152–58.

14. S. J. Nelson and M. E. Johnson, "Jens Munk Archipelago: Ordovician-Silurian Islands in the Churchill Area of the Hudson Bay Lowlands, Northern Manitoba," *Journal of Geology* 110 (2002): 577–89.

15. D. M. Rudkin, G. A. Young, and G. S. Nowlan, "The Oldest Horseshoe Crab: A New Xiphosurid from Late Ordovician Konservat-Lagerstätten Deposits, Manitoba, Canada," *Palaeontology* 51 (2008): 1–9.

16. Nelson and Johnson, "Jens Munk Archipelago."

17. The English translation from the original Danish of Captain Jen Munk's log is W. A. Kenyon, ed., *The Journal of Jens Munk, 1619–1620* (Toronto: Royal Ontario Museum, 1980).

18. Nelson and Johnson, "Jens Munk Archipelago."

19. M. E. Johnson and J.-Y. Rong, "Middle to Late Ordovician Rocky Bottoms and Rocky Shores from the Manitoulin Island Area (Ontario)," *Canadian Journal Earth Sciences* 26 (1989): 642–53.

20. An Upper Ordovician island from New South Wales in eastern Australia is described in I. G. Percival and B. D. Webby, "Island Benthic Assemblages: With Examples from the Late Ordovician of Eastern Australia," *Historical Biology* 11 (1996): 171–85.

4. HOW ISLANDS RECALL WINDWARD SURF AND LEEWARD CALM: A JOURNEY IN LATE SILURIAN TIME TO INNER MONGOLIA'S BATER ISLAND

1.   A. W. Grabau, *The Rhythm of the Ages* (Peking: Henry Vetch, 1940), 561.

2.   Contributions by Amadeus William Grabau to an early understanding of global sea-level changes and aspects of the geologist's career in China are outlined in M. E. Johnson, "A. W. Grabau's Embryonic Sequence Stratigraphy and Eustatic Curve," in *Eustasy: The Historical Ups and Downs of a Major Geological Concept*, ed. R. H. Dott Jr., (Boulder, CO: Geological Society of America, 1992), 111.

3.   J. Rong et al., "Continental Island from the Upper Silurian (Ludlow) Sino-Korean Plate," *Chinese Science Bulletin* 46 (2001): 238–41.

4.   M. E. Johnson, J.-Y. Rong, and Su Wen-bo, "Paleogeographic Orientation of the Sino-Korean Plate Based on Evidence for a Prevailing Silurian Wind Field," *Journal of Geology* 112 (2004): 671–84.

5.   C. M. Ø. Rasmussen et al., "Cascading Trend of Early Paleozoic Marine Radiations Paused by Late Ordovician Extinctions," *Proceedings of the National Academy of Sciences* 116, no. 15 (2019): 7207–13.

6.   J. Rong et al., "Coral-Stromatoporoid Faunas from the Shores of a Late Silurian Island, Inner Mongolia, North China," *Memoirs of the Association of Australasian Palaeontologists* 44 (2013): 95–105.

7.   Approved by the IUGS in 1980, the base of the Ludlow series is now defined by a Global Stratotype Section and Point (GSSP) at the first occurrence of a graptolite species (*Neodiversongraptus nilssoni*) within shale beds in the vicinity of Ludlow Castle in Shropshire, England. In turn, the Ludlow series is subdivided into two parts named the Gorstian stage (lower in position) and the Ludfordian stage (upper in position), both of which are formally defined by GSSPs Shropshire strata.

8.   M. E. Johnson et al., "Continental Island from the Upper Silurian (Ludfordian Stage) of Inner Mongolia: Implications for Eustasy and Paleogeography," *Geology* 29 (2001): 955–58.

9.   Rong et al., "Coral-Stromatoporoid Faunas."

10. Rong et al., "Coral-Stromatoporoid Faunas."

11. Johnson, Rong, and Su, "Paleogeographic Orientation of the Sino-Korean Plate."

12. The life and achievements of Temuchin (1162–1227), better known as Genghis Khan, are traced in the biography by F. McLynn, *Genghis Khan: His Conquests, His Empire, His Legacy* (Boston: Da Capo Press, 2015), 699.

13.   Stratigraphic and paleontological details from the upper Hemese beds at Kuppen correlated with the
       Ludfordian stage are described by M. Keeling and S. Kershaw, "Rocky Shore Environments in the
       Upper Silurian of Gotland, Sweden," *GFF* 116, no. 2 (1994): 69–74.

14.   L. Cherns, "Silurian Polyplacophoran Molluscs from Gotland, Sweden," *Palaeontology* 41 (1998):
       939–974.

15.   The Cambrian and Ordovician lineage of chitons is laid out in the study by B. Runnegar et al., "New
       Species of the Cambrian and Ordovician Chitons *Matthevia* and *Cheloides* from Wisconsin and
       Queensland: Evidence for the Early History of Polyplacophoran Mollusks," *Journal of Paleontology*
       53 (1979): 1374–94.

16.   C. M. Soja, "Potential Contributions of Ancient Oceanic Islands to Evolutionary Theory," *Journal
       of Geology* 100 (1992): 125–34.

5. HOW BIGGER ISLANDS ARE BROKEN INTO SMALLER PIECES: A JOURNEY IN LATE DEVONIAN TIME TO WESTERN AUSTRALIA'S MOWANBINI ARCHIPELAGO

1.   Based on dating from fossil graptolites that bracket strata containing the earliest *Baragwanathia*
       flora, the case is made that the oldest representatives appeared in Australia during the preceding
       Late Silurian according to R. B. Rickards, "The Age of the Earliest Club Mosses: The Silurian Barag-
       wanathia Flora in Victoria, Australia," *Geological Magazine* 137 (2000): 207–209.

2.   The extensive Devonian record of fossil fish from Australia is covered by C. J. Burrow and S. Turner,
       "Fossil Fish Taphonomy and the Contribution of Microfossils in Documenting Devonian Vertebrate
       History," In *Earth and Life*, ed. J. A. Talent (Dordrecht: Springer Science + Business Media, 2012),
       1100.

3.   P. E. Playford, R. M. Hocking, and A. E. Cockbain, "Devonian Reef Complexes of the Canning
       Basin, Western Australia," *Geological Survey of Western Australia Bulletin* 145 (2009): 403.

4.   M. E. Johnson and G. E. Webb, "Outer Rocky Shores of the Mowanbini Archipelago, Devonian
       Reef Complex, Canning Basin, Western Australia," *Journal of Geology* 115 (2007): 583–600.

5.   W. Buggisch, "The Global Frasnian-Famennian Kellwasser Event," *Geologische Rundschau* 80 (1991):
       49–72.

6.   M. E., Johnson et al., "Upper Devonian Shoal-Water Delta Integrated with Cyclic Back-Reef Facies
       off the Mowanbini Archipelago (Canning Basin), Western Australia," *Facies* 59 (2013): 991–1009.

7.   Johnson et al., "Upper Devonian Shoal-Water Delta."

8.   Johnson et al., "Upper Devonian Shoal-Water Delta."

9.   Johnson and Webb, "Outer Rocky Shores of the Mowanbini Archipelago."

10.   Johnson and Webb, "Outer Rocky Shores of the Mowanbini Archipelago."

11.   The basal boundary of the Frasnian stage is pinned to the first appearance of a unique cono-
       dont species (*Ancyrodella rotundilloba*). The top of the Frasnian stage corresponds to the base
       of the overlying Famennian stage with its GSSP also located in the Noire Mountains of France.
       That stage takes its name from the Famenne region of Belgium. The first appearance of another
       unique conodont (*Palmatolepis triangularis*) is used to distinguish the basal boundary of the
       Famennian stage. The lapse of time between the start of the Frasnian age and its conclusion is
       roughly ten million years, estimated to correspond to a phase in geologic time approximately 385
       million years ago.

12.   Playford, Hocking, and Cockbain, "Devonian Reef Complexes."

13.   Playford, Hocking, and Cockbain, "Devonian Reef Complexes."

14.   Playford, Hocking, and Cockbain, "Devonian Reef Complexes."

15. B. G. Baarli et al., "Shoal-Water Dynamics and Coastal Biozones in a Sheltered-Island Setting: Upper Devonian Pillara Limestone (Western Australia)," *Lethaia* 49 (2016): 507–23.

16. Baarli et al., "Shoal-Water Dynamics and Coastal Biozones."

17. Baarli et al., "Shoal-Water Dynamics and Coastal Biozones."

18. Playford, Hocking, and Cockbain, "Devonian Reef Complexes."

19. Johnson et al., "Upper Devonian Shoal-Water Delta."

20. The term for dendritic streams takes its name from the Greek word *dendron* for "tree," in allusion to the type of branching typical for the way a tree expands with upward growth. This kind of drainage is characteristic of erosion on a uniform ground surface like the relatively flat-lying rocks of the Devonian Pillara limestone.

21. The term for stream drainage in a trellis pattern derives from the Latin word *trilix*, in reference to a pattern of weaving with threads interwoven at right angles as in a screen. Folded strata exposed at the earth's surface are particularly susceptible to this kind of drainage, where some layers are more resistant to erosion parallel to softer rock layers more easily worn away by running water.

22. The post-Devonian history of the Canning basin is treated in detail by P. E. Playford, "Palaeokarst, Pseudokarst, and Sequence Stratigraphy in Devonian Reef Complexes of the Canning Basin, Western Australia," in *The Sedimentary Basins of Western Australia 3*, ed. M. Keep and S. J. Moss, 763–93 (Perth, Western Australia: Petroleum Exploration Society of Australia, 2002).

23. The essential facts on population statistics and human density patterns for Australia are regularly updated at https://www.worldometers.info/world-population/australia-population/.

24. The extraordinary story of song lines passed down orally through generations of Australia's earliest inhabitants is told with compassion by Bruce Chatwin, *The Songlines* (New York: Penguin, 1987), 294.

25. N. R. Fischbuch, "Devonian Reef-Building Stromatoporoids from Western Canada," *Journal of Paleontology* 44 (1970): 1071–84.

26. H. H. Tsien, "Paleoecology of Algal-Bearing Facies in the Devonian (Couvinian to Frasnian) Reef Complexes of Belgium," *Palaeogeography, Palaeoclimatology, Palaeoecology* 27 (1979): 103–127.

27. J.-W. Shen, G. E. Webb, and J. S. Jell, "Platform Margins, Reef Facies, and Microbial Carbonates: A Comparison of Devonian Reef Complexes in the Canning Basin, Western Australia, and the Guilin Region, South China," *Earth-Science Reviews* 88 (2008): 33–59.

6. HOW SOFTER ISLANDS DISSOLVE: A JOURNEY IN EARLY PERMIAN TIME TO THE LABYRINTH KARST OF WESTERN AUSTRALIA

1. For many years, I quoted this piece for shock value in my historical geology class at Williams College. Sharon Begley, "Ice Cubes for Penguins," *Newsweek*, April 2, 1995, 56.

2. The 14 pages in appendix 2 from T. Sunamura, *Geomorphology of Rocky Coasts* (Medford, MA: John Wiley, 1992), 302, lay out the most extensive set of published data on worldwide coastal cliff erosion rates.

3. The authoritative account of Western Australia's Carnarvon Basin is R. M. Hocking, H. T. Mores, and W. J. E. Van de Graaff, *Geology of the Carnarvon Basin Western Australia*, no. 133 (Perth: State Printing Division, 1987), 289.

4. Map sheets used to assemble a composite geologic picture for this chapter on the Gascoyne River area include Sheet SG/50-5 (1985) compiled by P. D. Denman and others, Sheet SG/50-1 compiled

by R. M. Hocking and others (1985), Sheet SG50-2 compiled by S. J. Williams and others (1983a), and sheet SG/50-6 compiled by S. G. Williams and others (1983b).

5. Hocking, Mores, and Van de Graaf, *Geology of the Carnarvon Basin*.

6. B. F. Glenister and W. M. Furnish, "The Permian Ammonoids of Australia," *Journal of Paleontology* 5 (1961): 673–736.

7. The unique conodont species that delimit the base of the Artinskian stage are *Sweetognathus whitei* and *Mesogondolella bisseli*. From the start, Russian territories played a commanding role in the advancement of terminology related to the Permian system. The Artinskian stage and corresponding age are named after the small Russian city Artinsk, located in the southern Ural Mountains. The Sakmarian stage and corresponding age are named in reference to strata on the Sakmara River, also located in the Ural Mountains. Agreement on global reference points (GSSPs) for these Permian units has yet to be ratified by responsible bodies under in the International Union of Geological Sciences.

8. The extent of the unique Permian bivalve fauna from eastern Australia is treated by B. Runnegar, "Ecology of *Eurydesma* and the *Eurydesma* Fauna, Permian of Eastern Australia," *Alcheringa* 3 (1979): 261–85.

9. M. E. Johnson and B. G. Baarli, "Development of Intertidal Biotas through Phanerozoic Time," in *Earth and Life*, ed. J. A. Talen (Dordrecht: Springer Science + Business Media, 2012), 1100.

10. N. J. Silberling, "Biogeographic Significance of the Upper Triassic Bivalve *Monotis* in Circum-Pacific Accreted Terranes: Tectonostratigraphic Terraces of the Circum-Pacific Region," *Earth Science Series*, no. 1, 1985 (2008): 63–70.

7. HOW ISLANDS REACT TO BIG STORMS: A JOURNEY IN EARLY JURASSIC TIME TO SAINT DAVID'S ARCHIPELAGO OF WALES

1. The statement by Charles Darwin on gaps in the rock record occurs in the first edition of *On the Origin of Species* (London: John Murray, 1859), chapter 9. "On the Imperfection of the Geological Record," 310–11.

2. The Beaufort scale was devised by British Royal Navy officer Francis Beaufort (1774–1857) in 1805 in an attempt to standardize wind and weather observations made at sea. The system was first used in an official capacity during the voyages of the HMS *Beagle* under Captain Robert FitzRoy (1805–1865), the second voyage of which Charles Darwin took part on a circumnavigation of the world from 1831 to 1836. FitzRoy not only applied the Beaufort scale during his active sea career but later, in 1854, ran what became the British Meteorological Office charged with making weather forecasts.

3. M. E. Johnson and W. S. McKerrow, "The Sutton Stone: An Early Jurassic Rocky Shore Deposit in South Wales," *Palaeontology* 38 (1995): 529–41.

4. A site of special scientific interest for citation is available from the Countryside Council for Wales regarding the Vale of Glamorgan and its unique geology and plant life: https://naturalresources .wales/media/677239/sssi_0129_citation_en0001.pdf.

5. J. L. Payne and M. E. Clapham, "End-Permian Mass Extinction in the Oceans: An Ancient Analogy for the Twenty-First Century?" *Annual Review of Earth and Planetary Sciences* 40 (2012): 89–111.

6. Payne and Clapham, "End-Permian Mass Extinction in the Oceans?"; B. Schoene et al., "Correlating the End-Triassic Mass Extinction and Flood Basalt Volcanisms at the 100-ka Level," *Geology* 38 (2010): 387–90.

7. D. V. Ager, "A Reinterpretation of the Basal 'Littoral Lias' of the Vale of Glamorgan," *Proceedings of the Geologists' Association* 97 (1986): 29–35.

8. M. E. Johnson and B. G. Baarli, "Development of Intertidal Biotas through Phanerozoic Time," in *Earth and Life*, ed. J. A. Talen (Dordrecht: Springer Science + Business Media, 2012), 1100.

9. The true identity of these fossils as corals was revealed in M. J. Simms, C. T. S. Little, and B. R. Rosen, "Corals Not Serpulids: Mineralized Colonial Fossils in the Lower Jurassic Marginal Facies of South Wales," *Proceedings of the Geologists' Association* 113 (2002): 31–36.

10. C. J. N. Fletcher, "Tidal Erosion, Solution Cavities and Exhalative Mineralization Associated with the Jurassic Unconformity at Ogmore, South Glamorgan," *Proceedings of the Geologists' Association* 99 (1988): 1–14.

11. A. E. Trueman, "The Liassic Rocks of Glamorgan," *Proceedings of Geologists' Association* 33 (1922): 245–84.

12. The stage name for the Hettangian derives from a town in northeastern France, Hettange-Grande, near the border with Luxembourg, where the characteristic ammonites from the first of the three zones (*Psiloceras planorbis*) are found in local strata. The formal golden spike for the Hettangian stage was approved in 2010 by the IUGS for a different place in Western Austria, although the name Hettangian has been conserved due to its extensive prior use.

13. Trueman, "The Liassic Rocks of Glamorgan."

14. Fletcher, "Tidal Erosion"; Johnson and McKerrow, "The Sutton Stone."; T. W. Sheppard, "Sequence Architecture of Ancient Rocky Shorelines and Their Response to Sea-Level Change: An Early Jurassic Example from South Wales, U.K.," *Journal of the Geological Society* 163 (2006): 595–606.

15. Fletcher, "Tidal Erosion"; Sheppard, "Sequence Architecture of Ancient Rocky Shorelines."

16. Ager, "A Reinterpretation of the Basal 'Littoral Lias.' "

17. Fletcher, "Tidal Erosion."

18. The synoptic history of the storm from September 16–30, 2017 is made by: R. J. Pasch, A. B. Penny, and R. Berg, "Hurricane Maria in the Atlantic Basin," *National Hurricane Center Tropical Cyclone Report: Hurricane Maria* (AL152017), National Oceanic and Atmospheric Administration and the National Weather Service (2018): 1–48.

19. Storm deposits left behind after the passage of Hurricane Donna in the Florida Keys in 1960 are described and illustrated by M. M. Ball, E. A. Shinn, and K. W. Stockman, "The Geologic Effects of Hurricane Donna in South Florida," *Journal of Geology* 75 (1967): 58397. Their work represents one of the first reports by qualified geologists to appraise modern sediments redeposited by a hurricane.

20. Early Jurassic (Hettangian) paleogeography for the British Isles and adjacent Europe is mapped by M. J. Bradshaw et al., "Jurassic," in *Atlas of Paleogeography and Lithofacies*, ed. J. C. W. Cope, J. K. Ingham, and P. F. Rawson (London: Geological Society, 1992), 153.

21. A. K. Bucheit and R. N. Donovan, "Lower Jurassic Borings and Fissures: A Rocky Shoreline Composed of Cambro-Ordovician Limestone, Isle of Skye, Scotland," *The Compass* 74 (1998): 47–54.

22. Bucheit and Donovan, "Lower Jurassic Borings and Fissures."

23. H. T. De la Beche, "On the Formation of the Rocks of South Wales and South Western England," *Memoirs of the Geological Survey of Great Britain* 1 (1846): 267.

24. Bucheit and Donovan, "Lower Jurassic Borings and Fissures."

25. See *Proceedings of the Bath Natural History and Antiquarian Field Club 1865–1866*, volume 1 (Bath, UK: Bleeck & Leech), 1–9.

26. K. M. Marsaglia and G. D. Klein, "The Paleogeography of Paleozoic and Mesozoic Storm Depositional Systems," *Journal of Geology* 91 (1983): 117–42.

8. HOW ISLAND LIFE ALIGNS WITH GLOBAL CURRENTS: A JOURNEY IN LATE CRETACEOUS TIME TO BAJA CALIFORNIA'S ERÉNDIRA ISLANDS

1. Running estimates for the percentage of rocky shores along any given segment of continental coast-lines around the world are tabulated by M. E. Johnson, "Why Are Ancient Rocky Shores So Uncommon?" *Journal of Geology* 96 (1988): 469–80.

2. A. A. Chiarenza et al., "Asteroid Impact, Not Volcanism, Caused the End-Cretaceous Dinosaur Extinction," *Proceedings of the National Academy of Sciences* 117 (2020): 17084–93.

3. Large-scale geological mapping of the entire northern state of Baja California was conducted by R. G. Gastil, R. P. Phillips, and E. C. Allison, "Reconnaissance Geologic Map of the State of Baja California," *Geological Society of America*, no. 3 (1973); maps at a scale 1: 250,000.

4. M. E. Johnson and M. L. Hayes, "Dichotomous Facies on a Late Cretaceous Rocky Island as Related to Wind and Wave Patterns (Baja California, Mexico)," *Palaios* 8 (1993): 385–95.

5. Johnson and Hayes, "Dichotomous Facies."

6. H. L. Lescinsky, J. Ledesma-Vázquez, and M. E. Johnson, "Dynamics of Late Cretaceous Rocky Shores (Rosario Formation) from Baja California Mexico," *Palaios* 6 (1991): 126–41.

7. Taking its name from exposures on a hillside called La Grande Champagne at Aubeterre-sur Dronne in northern France, the top of the Campanian stage is defined by the base of the overlying Maastrichtian stage. The Maastrichtian stage is the last of twelve such stages that compose the Cretaceous system. It derives its name from chalk beds exposed around the city of Maastricht in the Netherlands. However, the GSSP approved under authority of the International Union of Geological Sciences is sited in a quarry near the town of Dax in southwestern France, equated with the first appearance of a particular ammonite (*Pachydiscus neubergicus*).

8. Biostratigraphic details on the correlation of ammonite zones from the Upper Cretaceous of North America in conjunction with changes in magnetic signatures are treated by P. D. Ward, J. W. Haggart, and R. Mitchell, "Integration of Macrofossil Biostratigraphy and Magnetostratigraphy for the Pacific Coast Upper Cretaceous (Campanian-Maastrichtian) of North America and Implications for Correlation with the Western Interior and Tethys," *Geological Society of America Bulletin* 124 (2012): 957–74.

9. M. E. Johnson, J. Ledesma-Vázquez, and B. G. Baarli, "Vertebrate Remains on Ancient Rocky Shores: A Review with Report on Hadrosaur Bones from the Upper Cretaceous of Baja California (Mexico)," *Journal of Coastal Research* 22 (2006): 574–80.

10. M. E. Johnson, J. Ledesma-Vázquez, H.C. Clark, and J.A. Zwiebel, "Coastal Evolution of Late Cretaceous and Pleistocene Rocky Shores: Pacific Rim of Northern Baja California, Mexico," *Geological Society of America Bulletin* 108 (1996): 708–21.

11. Johnson et al., "Coastal Evolution of Late Cretaceous and Pleistocene Rocky Shores."

12. H. C. Clark and M. E. Johnson, "Coastal Geomorphology of Andesite from the Cretaceous Alisitos Formation in Baja California (Mexico)," *Journal of Coastal Research* 11 (1995): 401–14.

13. The significance of joint patterns and their spacing as a factor in the erosion of igneous rocks by wave action is treated by Clark and Johnson, "Coastal Geomorphology of Andesite."

14. Gastil, Phillips, and Allison, "Reconnaissance Geologic Map of the State of Baja California."

15. Johnson et al., "Coastal Evolution of Late Cretaceous and Pleistocene Rocky Shores."

16. Comparison of Charles Darwin's biological and geological observations during the 1831–1836 voyage of the HMS *Beagle* is documented regarding the number of days he spent on various islands and the number of pages he devoted to his earliest published observations. See the paper by M. E. Johnson and B. G. Baarli, "Charles Darwin in the Cape Verde and Galápagos Archipelagos: The Role of

Serendipity in Development of Theories on the Ups and Downs of Oceanic Islands," *Earth Sciences History* 4, no. 2 (2015): 220–42.

17. Dissections revealing the stomach contents of "minced seaweed" from several marine iguanas are described by C. Darwin, "Galapagos Archipelago," in *Journal of Researches into the Natural History and Geology of Countries Visited During the Voyage of the H.M.S.* Beagle *Round the World, Under the Command of Capt. Fitz Roy, R.N.*, 2nd ed. (London: John Murray, 1845), 520.

18. Johnson, Ledesma-Vázquez, and Baarli, "Vertebrate Remains on Ancient Rocky Shores."

19. W. J. Morris, "A New Species of Hadrosaurian Dinosaur from the Upper Cretaceous of Baja California: *Lambeosaurus laticaudus*," *Journal of Paleontology* 55 (1981): 453–62.

20. N. Ibrahim et al., "Tail-Propelled Aquatic Locomotion in a Theropod Dinosaur," *Nature* 581 (2020): 67–70; T. Beevor et al., "Taphonomic Evidence Supports an Aquatic Lifestyle for *Spinosuarus*," *Cretaceous Research* 117 (2021): 104627.

21. Reconstruction of middle Late Cretaceous oceanic circulation including divergent counter currents flowing northward on the western shore of North America and southward along the northwest shore of South America are inferred by W. A. Gordon, "Marine Life and Ocean Surface Currents in the Cretaceous," *Journal of Geology* 81 (1973): 269–84.

22. The dynamics of the Leeuwin Current off the west coast of Western Australia are described and compared to phenomenon that may have occurred in the Americas by A. J. Weaver, "Ocean Currents and Climate," *Nature* 374 (1990): 432.

23. R. Riosmena-Rodríguez, W. Nelson, and J. Aguirre, eds., *Rhodolith/Maërl Beds: A Global Perspective* (Cham, Switzerland: Springer, 2017), 368.

24. L. A. Buatois and A. Encinas, "Ichnology, Sequence Stratigraphy and Depositional Evolution of an Upper Cretaceous Rocky Shoreline in Central Chile: Bioerosion Structures in a Transgressed Metamorphic Basement," *Cretaceous Research* 32 (2011): 203–12.

25. J. S. Crampton, "A Late Cretaceous Near-Shore Rocky Substrate Macrofauna from Northern Hawkes Bay, New Zealand," *New Zealand Geological Survey Record* 35 (1988): 21–24.

26. F. Surlyk and W. K. Christensen, "Epifaunal Zonation on an Upper Cretaceous Rocky Coast," *Geology* 2 (1974): 529–34; F. Surlyk and A. M. Sørensen, "An Early Campanian Rocky Shore at Ivö Klack, Southern Sweden," *Cretaceous Research* 31 (2010): 567–76.

27. M. A. Wilson and P. D. Taylor, "Palaeoecology of Hard Substrate Faunas from the Cretaceous Quhlah Formation of the Oman Mountains," *Palaeontology* 44 (2001): 21–41.

9. HOW ISLAND LIFE ADJUSTS TO OPPOSING SHORES ON OCEANIC ISLANDS: A JOURNEY IN MIDDLE MIOCENE TIME TO THE MADEIRA ARCHIPELAGO

1. The age of discovery and the wealth it brought to Portugal and Spain are put into context with the history of geology by H. Leitão and W. Alvarez, "The Portuguese and Spanish Voyages of Discovery and the Early History of Geology," *Geological Society of America Bulletin* 123 (2011): 1219–33.

2. Fossil plants that include laurel leaves are described by C. Góis-Marques, J. Madeira, and M. Menezes de Sequeira, "Inventory and Review of the Mio-Pleistocene São Jorge Flora (Madeira Island, Portugal): Palaeoecological and Biogeographical Implications," *Journal of Systematic Palaeontology* 16, no. 2 (2017), 159–77, https://www.doi.org/10.1080/14772019.2017.1282991.

3. R. S. Ramalho et al., "The Emergence of Volcanic Oceanic Islands on a Slow-Moving Plate: The Example of Madeira Island, NE Atlantic," *Geochemistry Geophysics Geosystems* 16, no. 2 (2015), 522–37, https://www.doi.org/10.1002/2014GC005657.

4.   A. Santos et al., "Miocene Intertidal Zonation on a Volcanically Active Shoreline: Porto Santo in the Madeira Archipelago, Portugal," *Lethaia* 44 (2011): 26–32.

5.   Santos et al., "Miocene Intertidal Zonation."

6.   M. E. Johnson et al., "Rhodolith Transport and Immobilization on a Volcanically Active Rocky Shore: Middle Miocene at Cabeço das Laranjas on Ilhéu de Cima (Madeira Archipelago, Portugal)," *Palaeogeography, Palaeoclimatology, Palaeoecology* 300 (2011): 113–27.

7.   The life span of the largest rhodoliths is based on growth bands exposed in cuts through the algal thallus as described by R. Riosmena-Rodrígues et al., "Reefs that Rock and Roll: Biology and Conservation of Rhodolith Beds in the Gulf of California," in *The Gulf of California—Biodiversity and Conservation*, ed. R. Brusca, 49–71 (Tucson: University of Arizona Press, 2010).

8.   Johnson et al., "Rhodolith Transport and Immobilization."

9.   Rare hurricane strikes on Madeira are reviewed by J. M. Vaquero et al., "A Historical Analog of the 2005 Hurricane Vince," *American Meteorological Society* 85 (2008): 191–201.

10. A. G. Santos et al., "Basalt Mounds and Adjacent Depressions Attract Contrasting Biofacies on a Volcanically Active Middle Miocene Shoreline (Porto Santo, Madeira Archipelago, Portugal)," *Facies* 58 (2012a): 573–85.

11. Santos et al., "Basalt Mounds and Adjacent Depressions."

12. B. G. Baarli et al., "A Middle Miocene Carbonate Embankment on an Active Volcanic Slope: Ilhéu de Baixo, Madeira Archipelago, Eastern Atlantic," *Geological Journal* 49 (2014): 90–106.

13. A. Santos et al., "Miocene Intertidal Zonation on a Volcanically Active Shoreline."

14. A. Santos et al., "Symbiotic Association of a Pyrgomatid Barnacle and a Coral from a Volcanic Middle Miocene Shoreline (Porto Santo, Madeira Archipelago, Portugal)," *Palaeontology* 55 (2012b): 173–82; B. G. Baarli et al., "Miocene to Pleistocene Transatlantic Dispersal of *Ceratoconcha* Coral-Dwelling Barnacles and North Atlantic Island Biogeography," *Palaeogeography, Palaeoclimatology, Palaeoecology* 468 (2017): 520–28.

15. Studies on living and fossil barnacles occupied Charles Darwin for many years prior to his 1859 publication *On the Origin of Species*. The detailed taxonomic work on which he honed his descriptive skills is: *A Monograph on the Sub-Class Cirripedia with Figures of all the Species* (London: Ray Society Publication, 1854), 684.

16. Santos et al., "Symbiotic Association of a Pyrgomatid Barnacle"; Baarli et al., "Miocene to Pleistocene Transatlantic Dispersal."

10. HOW VOLCANIC ISLANDS RISE, FALL AND RENEW: A JOURNEY IN EARLY PLIOCENE TIME TO THE AZOREAN SANTA MARIA ISLAND

1.   Variables regarding volcano classification and styles are reviewed in greater detail by A. Scarth, *Volcanoes: An Introduction* (London: University College Press Limited, 1994), 273.

2.   R. S. Ramalho et al., "Coastal Evolution on Volcanic Oceanic Islands: A Complex Interplay between Volcanism, Erosion, Sedimentation, Sea Level Change and Biogenic Production," *Earth-Science Reviews* 127 (2013): 140–70.

3.   The diversity of Pliocene and Pleistocene fossils from this island for both invertebrate and vertebrate faunas is reviewed by S. P. Ávila et al., "The Marine Fossil Record at Santa Maria Island (Azores)," in *Volcanoes of the Azores*, ed. U. Kueppers and C. Beier (Berlin: Springer-Verlag, 2018), 355.

4.   R. S. Ramalho et al., "Emergence and Evolution of Santa Maria Island (Azores)—The Conundrum of Uplifted Islands Revisited," *Geological Society of America Bulletin* 129 (2017): 372–91.

5.   The Figueiral mine portal is illustrated by S. P. Ávila et al., "The Marine Fossil Record."

6.   Recognized by the IUGS in 2000, the Zanclean stage is the lower of two stages that define the Pliocene. The name derives from Zancle, the pre-Roman name for the city of Messina on Sicily. It may seem odd to invoke the geology of a place far away in Italy when describing the geology of an Azorean island, but the formal nomenclature of the IUGS is essential where detailed correlations of strata on a global basis are required to understand what happened elsewhere in the Pliocene world when Santa Maria Island was being formed. In this case, the base of the Zanclean stage is defined by multiple means, including the top of a geomagnetic marker (Cr3) and the first appearance of a species of nanoplankton (*Ceratolithus acutus*). While that specific nanoplankton may or may not be present in marine sedimentary rocks from the Touril complex, it is expected that the magnetic marker should be present in the underlying igneous rocks. Technically, the top of the Zanclean stage is defined by the base of the overlying Piacenzian stage, which is named for the city of Piacenza on the Po River in northern Italy. That boundary is defined by rocks at the base of another geomagnetic marker (C2An).

7.   Ramalho et al., "Emergence and Evolution of Santa Maria Island (Azores)."

8.   Perturbations of the global Pliocene Warm Period are defined and quantified by C. M. N. Brierly et al., "Greatly Expanded Tropical Warm Pool and Weakened Hadley Circulation in the Early Pliocene," *Science* 323, no. 5922 (2009): 1714–18.

9.   Fossil traces of rock-boring echinoids newly recognized to science are described from the locality at Ichnofossils' Cave by A. Santos et al., "Role of Environmental Change in Rock-Boring Echinoid Trace Fossils," *Palaeogeography, Palaeoclimatology, Palaeoecology* 432 (2015): 1–14.

10. A. C. Rebelo et al., "Rocking Around a Volcanic Island Shelf: Pliocene Rhodolith Beds from Malbusca, Santa Maria Island (Azores, NE Atlantic)," *Facies* 62 (2016): 1–31.

11. A. Uchman et al., "Feeding Traces of Recent Ray Fish and Abundant Occurrences of the Trace Fossil *Piscichnus waitemata* from the Pliocene of Santa Maria Island Azores (Northeast Atlantic)," *Palaios* 33 (2018): 361–75.

12. M. E. Johnson et al., "Intense Hurricane Transports Sand Onshore: Example from the Pliocene Malbusca Section on Santa Maria Island (Azores, Portugal)," *Marine Geology* 385 (2017): 244–94.

13. Johnson et al., "Intense Hurricane Transports."

14. S. P. Ávila et al., "*Persististrombus coronatus* (Mollusca: Strombidae) in the Lower Pliocene of Santa Maria Island (Azores, NE Atlantic): Paleoecology, Paleoclimatology and Paleobiogeographic Implications," *Palaeogeography, Palaeoclimatology, Palaeoecology* 441 (2016): 912–23.

15. S. P. Ávila et al., "Palaeoecology, Taphonomy, and Preservation of a Lower Pliocene Shell Bed (Coquina) from a Volcanic Oceanic Island (Santa Maria Island, Azores, NE Atlantic Ocean)," *Palaeogeography, Palaeoclimatology, Palaeoecology* 430 (2015): 57–73.

16. A. Uchman et al., "Vertically-Oriented Trace Fossil *Macaronichnus segregatis* from Neogene of Santa Maria Island (Azores; NE Atlantic) Records Vertical Fluctuations of the Coastal Groundwater Mixing Zone on a Small Oceanic Island," *Geobios* 49 (2016): 229–41.

17. Ramalho et al., "Emergence and Evolution."

18. R. P. Meireles et al., "Depositional Processes on Oceanic Island Shelves—Evidence from Storm-Generated Neogene Deposits from the Mid-North Atlantic," *Sedimentology* 60 (2013): 1769–85.

19. Ramalho et al., "Emergence and Evolution."

20. A. Uchman et al. 2020. "Neogene Marine Sediments and Biota Encapsulated Between Lava Flows on Santa Maria Island (Azores, North-East Atlantic): An Interplay Between Sedimentary Erosional and Volcanic Processes." *Sedimentology* 67: 3595–3618.

21. Ramalho et al., "Emergence and Evolution."

22. M. E. Johnson et al., "Intense Hurricane Transports."

23. Frequency of storm events in the Azores is analyzed in: C. Andrade et al., "Comparing Historic Records of Storm Frequency and the North Atlantic Oscillation (NAO) Chronology for the Azores Region," *Holocene* 18, no. 5 (2008): 745–54.

24. Ramalho, personal communication.

25. A roadside turnout provides access to the small shelter high above Ponte do Castelo, where lookouts were posted on watch for whales. The site features signage with information regarding the whaling station operated by the Companhia Baleeira Mariense as late as 1966.

26. The Pliocene (Zanclean) paleoislands of the Santa Ines Archipelago are described in the guidebook by M. E. Johnson, *Discovering the Geology of Baja California: Six Hikes on the Southern Gulf Coast* (Tucson: University of Arizona Press, 2002), 220.

27. Migration of a distinctive bryozoan species from the Atlantic Ocean to the Pacific Ocean through the Pliocene straits of Panama is documented by R. Coffey and M. E. Johnson, "Bryozoan Nudules Built around Andesite Clasts from the Upper Pliocene of Baja California: Paleoecological Implications and Closure of the Panama Isthmus," in *Pliocene Carbonates and Related Facies Flanking the Gulf of California*, ed. M. E. Johnson and J. Ledesma-Vázquez (Mexico: Baja California, Geological Society of America Special Paper 318, 1997), 171.

28. A. J. Weaver, "Ocean Currents and Climate," *Nature* 374 (1990): 432. See endnotes for chapter 8.

29. Brierley et al., "Greatly Expanded Tropical Warm Pool and Weakened Hadley Circulation in the Early Pliocene." *Science* 323, no. 5922 (2009): 1714–18.

30. Ávila et al., "*Persististrombus coronatus* (Mollusca: Strombidae)."

11. HOW THE YOUNGEST ISLANDS CHALLENGE WITNESS: JOURNEYS IN PLEISTOCENE TIME TO ISLANDS ON THE AFRICAN AND PACIFIC TECTONIC PLATES

1. The geology of the Praia harbor area and southeast coast of Santiago Island in the Cape Verde Archipelago (including descriptions of fossil rhodoliths) forms the first chapter in: C. Darwin, *Geological Observations on the Volcanic Islands Visited During the Voyage of the H.M.S. Beagle* (London: Smith, Elder & Co., 1844), 175.

2. M. E. Johnson et al., "Rhodoliths, Uniformitarianism, and Darwin: Pleistocene and Recent Carbonate Deposits in the Cape Verde and Canary Archipelagos," *Palaeogeography, Palaeoclimatology, Palaeoecology* 329–330 (2012): 83–100.

3. Intended only for his immediate family, Darwin's recollections were first published by his granddaughter, Nora Barlow, *The Autobiography of Charles Darwin 1809–1882* (with the Original Omissions Restored, Edited and With Appendix and Notes by His Grand-daughter, Nora Barlow) (London: Collins, 1958), 253.

4. Baarli et al., "What Darwin Did Not See: Pleistocene Fossil Assemblages on a High-Energy Coast at Ponta das Bicudas, Santiago Cape Verde Islands," *Geological Magazine* 150 (2013): 183–89.

5. E. Mayoral et al., "Upper Pleistocene Trace Fossils from Ponta das Bicudas, Santiago, Cape Verde Islands: Systematics, Taphonomy and Palaeoenvironmental Evolution," *Palaeogeography, Palaeoclimatology, Palaeoecology* 498 (2018): 83–98.

6. Johnson et al., "Rhodoliths, Uniformitarianism, and Darwin."

7. The cross-section drafted by Charles Darwin showing the stretch of coastline below the small volcano at "Signal Post Hill" is found on Charles Darwin, *Geological Observations*, 9.

8. M. E. Johnson and B. G. Baarli, "Erosion and Burial of Granite Rocky Shores in the Recent and Late Pleistocene of the Seychelles Islands: Physical and Biological Perspectives," *Journal of Coastal Research* 21 (2005): 867–79.

9. Johnson and Baarli, "Erosion and Burial of Granite Rocky Shores."

10. Johnson and Baarli, "Erosion and Burial of Granite Rocky Shores."

11. Johnson and Baarli, "Erosion and Burial of Granite Rocky Shores."

12.   M. E. Johnson et al., "On the Rise and Fall of Oceanic Islands: Towards a Global Theory Following from the Pioneering Studies of Charles Darwin and James Dwight Dana," *Earth-Sciences Review* 180 (2018): 17–36.

13.   Dana's official report of the geology of the Wilkes expedition was published in "Geology," in *United States Exploring Expedition During the Years 1838, 1839, 1840, 1841*, ed. C. Wilkes (New York: George Putnam, 1849), 398.

14.   C. Darwin, *The Structure and Distribution of Coral Reefs, Being the First Part of the Geology of the Voyage of the Beagle, Under the Command of Capt. FitzRoy, RN during the Years 1832–1836* (London: Smith Elder & Co., 1842), 214.

15.   Johnson et al., "On the Rise and Fall of Oceanic Islands."

16.   R. H. MacArthur and E. O. Wilson, *The Theory of Island Biogeography* (Princeton, NJ: Princeton University Press, 1967), 203.

17.   J. D. Dana, *Corals and Coral Islands* (New York: Dodd and Mead, 1872), 398.

18.   C. Darwin, *A Monograph on the Sub-Class Cirripedia with Figures of all the Species* (London: Ray Society Publication, 1854).

19.   Experiments on viability of plant seeds after submersion in saltwater that Charles Darwin ran with the help of his children at Down House are reviewed by J. T. Costa, "The Impish Side of Evolution's Icon," *American Scientist* 106 (2018): 104–11.

20.   Johnson et al., "On the Rise and Fall of Oceanic Islands."

12. HOW ISLANDS DRAW MEANING AND OBLIGATION: DESCENDING MOUNT MISEN ON JAPAN'S SACRED MIYAJIMA

1.   R. S. White and J. K. Greenberg, "Religious Faith as a Motivation in Using Geosciences to Develop a Sustainable Future," in *Geoscience for the Public Good and Global Development: Toward a Sustainable Future*, ed. G. R. Wessel and J. K. Greeenberg (Boulder, CO: Geological Society of America, 2016), 478.

2.   W. M. Davis, "The Physical Geography of Southern New England," *National Geographic Monograph, National Geographic Society* 1 (1895): 269–83.

3.   "The Life and Legend of Kobo Daishai (Kukai)" is recounted in chapter 3 of P. L. Nicoloff, *Sacred Koyasan* (Albany: State University of New York, 2008), 392.

4.   A photo essay on Shikoku and the island's famous pilgrimage route appears in D. George, "On the Pilgrim's Path," *Islands* 21 (2001): 82–95.

5.   Adherents of Shingon Buddhism maintain a highly personal relationship to the Kobo Daishai, founder in 796 of the Kyoto temple Kyo-o-gokoku-ji (known more informally as the temple Toji). Each morning at 6:00 A.M., the compound gates open, and worshipers enter a small hall after the strike of ten bells. There, a devotional service is performed by three monks, one of who offers food before a ninth-century statue of Kukai. The author attended such a service in March 2019 and was rewarded with Kukai's blessing. The temple's five-story pagoda is the tallest in Japan. It is a copy of the original rebuilt in 1644 based on an extant Chinese model Kukai would have been familiar with in Xi'an.

6.   Published posthumously in 1739, an English translation from the original Norwegian text is available as P. Dass, *The Trumpet of Nordland* (Helgeland, Norway: Helgeland Museum, 2015), 175.

7.   The story of the Oglala Sioux sage is told by J. G. Neihardt, *Black Elk Speaks* (Lincoln: University of Nebraska Press, 1979), 230.

8.   C. Lochman-Balk, "Upper Cambrian Faunal Patterns on the Craton," *Geological Society of America Bulletin* 81 (1970): 3197–224.

9.   Black Hills ecology is covered by E. Raventon, *Island in the Plains, A Black Hills Natural History* (Boulder, CO: Johnson, 1994), 272.

10. The author was granted permission to climb Dreamer's Rock during the summer of 1987 following fieldwork on Canada's Manitoulin Island. The summit is photographed (fig. 2A) in M. E. Johnson and J.-Y. Rong, "Middle to Late Ordovician Rocky Bottoms and Rocky Shores from the Manitoulin Island Area (Ontario)," *Canadian Journal Earth Sci*ences 26 (1989): 642–53.

11. Essays in the tradition of Rachel Carson celebrate the natural history of Acadia National Park, as found in C. Camuto, *Time and Tide in Acadia* (Woodstock, VT: Countryman, 2009), 199.

12. D. R. Muhs and L. T. Groves, "Little Islands Recording Global Events: Late Quaternary Sea Level History and Paleozoogeography of Santa Barbara and Anacapa Islands, Channel Islands National Park, California," *Western North American Naturalist* 74 (2018): 540–89.

13. A. von Humboldt, *Personal Narrative of Travels to the Equinoctial Regions of the New Continent During the Years 1799–1804* (London: Longman, Hurst, Rees, Orme, and Brown, 1814), 386.

14. Data on the growth of visitor pressures on Galapagos National Park are summarized in the article by V. Toral-Granda, "Stowaways to Paradise," *Galapagos News* (Spring–Summer 2018): 9.

15. M. E. Johnson and J. Ledesma-Vázquez, *Gulf of California Coastal Ecology: Insights from the Present and Patterns from the Past* (San Diego: Sunbelt, 2016), 134.

16. A popular account surrounding the controversy of Darwin's geomorphic model of atoll development and its final resolution with the drilling program on Eniwetok prior to the testing of atomic bombs is provided by D. Dobbs, *Reef Madness: Charles Darwin, Alexander Agassiz, and the Meaning of Coral* (New York: Pantheon, 2005), 306.

17. Johnson and Ledesma-Vázquez, *Gulf of California Coastal Ecology*.

18. During expert testimony to the U.S. Congress in 1988, NASA scientist James Hansen predicted that Planet Earth would warm by an additional 1.9°F. Thirty years later, according to NASA measurements, Earth's average surface temperature had risen 1.6°F in confirmation of Hanson's testimony. Newspapers around the United States widely reported the finding at the time of the anniversary. It appeared in my local paper as: S. Borenstein and N. Forster, "A Warming Warning Come True," *Berkshire Eagle*, June 19, 2018. Soon after, an editorial by A. Revkin under the title "Climate Change First Became News 30 Years Ago. Why Haven't We Fixed It?" appeared in *National Geographic* 234 (July 2018): 17–20.

19. USGCRP, *Impacts, Risks, and Adaptation in the United States: Fourth National Climate Assessment, Volume II: Report-in Brief*, ed. D. R. Reidmiller et al. (Washington, DC: Publication of the US Global Change Program, 2018), 186.

20. United Nations Office For Disaster Risk Reduction, "Our World at Risk: Transforming Governance for a Resilient Future" (Geneva, Switzerland: UNDRR), 18, http://www.undrr.org/GAR2022.

21. P. Dukes, *Minutes to Midnight: History and the Anthropocene Era from 1763* (London, Anthem, 2011), 166.

BIBLIOGRAPHY

Ager, D. V. 1986. "A Reinterpretation of the Basal 'Littoral Lias' of the Vale of Glamorgan." *Proceedings of the Geologists' Association* 97, no. 1: 29–35.

Ali, J. R. 2017. "Islands as Biological Substrates: Classification of the Biological Assemblage Components and the Physical Island Types." *Journal of Biogeography* 44, no. 5 (May 2017): 984–94. https://doi.org/10.1111/bji.12872.

Andrade, C., R. M. Trigo, M. C. Freitas, M. C. Gallego, P. Borges, and A. M. Ramos. 2008. "Comparing Historic Records of Storm Frequency and the North Atlantic Oscillation (NAO) Chronology for the Azores Region." *The Holocene* 18, no. 5 (August 1, 2008):

Ávila, S. P., C. Melo, B. Berning, R. Cordeiro, B. Landau, and C. M. da Silva. 2016. "*Persististrombus coronatus* (Mollusca: Strombidae) in the Lower Pliocene of Santa Maria Island (Azores, NE Atlantic): Paleoecology, Paleoclimatology and Paleobiogeographic Implications." *Palaeogeography, Palaeoclimatology, Palaeoecology* 441, no. 4 (January 2016): 912–23.

Ávila, S. P., R. S. Ramalho, J. M. Habermann, R. Quartau, A. Kroh, B. Berning, M. Johnson, et al. 2015. "Palaeoecology, Taphonomy, and Preservation of a Lower Pliocene Shell Bed (Coquina) from a Volcanic Oceanic Island (Santa Maria Island, Azores, NE Atlantic Ocean)." *Palaeogeography, Palaeoclimatology, Palaeoecology* 430, no. 4: 57–73.

Ávila, S. P., R. S. Ramalho, J. M. Haberman, and J. Titschack. 2018. "The Marine Fossil Record at Santa Maria Island (Azores)." In *Volcanoes of the Azores*, ed. U. Kueppers and C. Beier, 155–96. Berlin: Springer-Verlag.

Baarli, B. G., M. Cachão, C. M. da Silva, M. E. Johnson, E. J. Mayoral, and A. Santos. 2014. "A Middle Miocene Carbonate Embankment on an Active Volcanic Slope: Ilhéu de Baixo, Madeira Archipelago, Eastern Atlantic." *Geological Journal* 49, no. 1 (January/February 2014): 90–106.

Baarli, B. G., M. C. D. Malay, A. Santos, M. E. Johnson, C. M. da Silva, J. Meco, M. Cachão, E. J. Mayoral. 2017. "Miocene to Pleistocene Transatlantic Dispersal of *Ceratoconcha* Coral-Dwelling Barnacles and North Atlantic Island Biogeography." *Palaeogeography, Palaeoclimatology, Palaeoecology* 468, no. 15 (February 2017): 520–28.

Baarli, B. G., A. G. Santos, E. J. Mayoral, J. Ledesma-Vázquez, M. E. Johnson, C. da Silva, and M. Cachão. 2013. "What Darwin Did Not See: Pleistocene Fossil Assemblages on a High-Energy Coast at Ponta das Bicudas, Santiago, Cape Verde Islands. *Geological Magazine* 150: 183–89.

Baarli, B. G., G. Webb, M. E. Johnson, A. G. Cook, and D. R. Walsh. 2016. "Shoal-Water Dynamics and Coastal Biozones in a Sheltered-Island Setting: Upper Devonian Pillara Limestone (Western Australia)." *Lethaia* 49: 507–523.

Ball, M. M., E. A. Shinn, and K. W. Stockman. 1967. "The Geologic Effects of Hurricane Donna in South Florida." *Journal of Geology* 75: 583–97.

Barlow, N. 1958. *The Autobiography of Charles Darwin 1809–1882*. London: Collins.

Beevor T., A. Quigley, R. E. Smith, R. S. H. Smyth, N. Ibrhiam, S. Zouhril, and D. M. Martill. 2021. "Taphonomic Evidence Supports an Aquatic Lifestyle for *Spinosuarus*." *Cretaceous Research* 117.

Bradshaw, M. J., J. C. W. Cope, D. W. Cripps, D. T. Donovan, M. K. Howarth, P. F. Rawson, I. M. West, and W. A. Wimbledon. 1992. "Jurassic." In *Atlas of Paleogeography and Lithofacies*, ed. J. C. W. Cope, J. K. Ingham, and P. F. Rawson. London: Geological Society.

Brierly, C. M., A. V. Fedorov, Z. Liu, T. D. Herbert, K. T. Lawrence, and J. P. Lariviere. 2009. "Greatly Expanded Tropical Warm Pool and Weakened Hadley Circulation in the Early Pliocene." *Science* 323, no. 5922: 1714–18.

Briggs, D., E. Clarkson, and R. Aldridge. 1983. "The Conodont Animal." *Lethaea* 16: 1–14.

Buatois, L.A., and A. Encinas. 2011. "Ichnology, Sequence Stratigraphy and Depositional Evolution of an Upper Cretaceous Rocky Shoreline in Central Chile: Bioerosion Structures in a Transgressed Metamorphic Basement." *Cretaceous Research* 32: 203–12.

Bucheit, A.K., and R. N. Donovan. 1998. "Lower Jurassic Borings and Fissures: A Rocky Shoreline Composed of Cambro-Ordovician Limestone, Isle of Skye, Scotland." *Compass* 74: 47–54.

Buggisch, W. 1991. "The Global Frasnian-Famennian Kellwasser Event." *Geologische Rundschau* 80: 49–72.

Burrow, C. J., and S. Turner. 2012. "Fossil Fish Taphonomy and the Contribution of Microfossils in Documenting Devonian Vertebrate History." In *Earth and Life*, ed. J. A. Talent, 1100. Dordrecht: Springer Science + Business Media.

Camuto, C. 2009. *Time and Tide in Acadia*. Woodstock, VT: Countryman.

Chatwin, B. 1987. *The Songlines*. New York: Penguin.

Cherns, L. 1998. "Silurian Polyplacohoran Molluscs from Gotland, Sweden." *Palaentology 41: 939–974.*

Chiarenza, A. A., A. Farnsworth, P. D. Mannion, D. J. Lunt, P. J. Valdes, J. V. Morgan, and P. A. Allison. 2020. "Asteroid Impact, Not Volcanism, Caused the End-Cretaceous Dinosaur Extinction." *Proceedings of the National Academy of Sciences* 117: 17084–93.

Clark, H.C., and M. E. Johnson. 1995. "Coastal Geomorphology of Andesite from the Cretaceous Alisitos Formation in Baja California (Mexico)." *Journal of Coastal Research* 11: 401–14.

Costa, J. T. 2018. "The Impish Side of Evolution's Icon." *American Scientist* 106: 104–11

Crampton, J. S. 1988. "A Late Cretaceous Near-Shore Rocky Substrate Macrofauna from Northern Hawkes Bay, New Zealand." *New Zealand Geological Survey Record* 35: 21–24.

Dalziel, I. W. D. and R. H. Dott, Jr. 1970. Geology of the Baraboo District Wisconsin, *Geological and Natural History Survey Information Circular Number 14*, 163 p.

Dana, J. D. *Corals and Coral Islands*. New York: Dodd and Mead, 1872.

Darwin, C. "Galapagos Archipelago." 1842. *The Structure and Distribution of Coral Reefs, Being the First Part of the Geology of the Voyage of the* Beagle, *Under the Command of Capt. FitzRoy, RN During the Years 1832–1836*. London: Smith Elder.

——. 1844. *Geological Observations on the Volcanic Islands Visited During the Voyage of the H.M.S. Beagle*. London: Smith, Elder.

——. 1845. In *Journal of Researches into the Natural History and Geology of Countries Visited During the Voyage of the H.M.S. Beagle Round the World, Under the Command of Capt. Fitz Roy, R.N.* 2nd ed. London: John Murray.

——. 1859. *On the Origin of Species*. London: John Murray.

Dass, P. 2015. *The Trumpet of Nordland*. Helgeland, Norway: Helgeland Museum.

Davis, R. A. 2006. "Precambrian Tidalites from the Baraboo Quartzite District, Wisconsin." *Marine Geology* 235, no. 1 (December 2006): 247–53.

Davis, W. M. 1895. "The Physical Geography of Southern New England." *National Geographic Monograph, National Geographic Society* 1: 269–83.

De la Beche, H. T. 1846. "On the Formation of the Rocks of South Wales and South Western England." *Memoirs of the Geological Survey of Great Britain* 1: 267.

Demski, M. W., B. J. Wheadon, L. A. Stewart, R. J. Elias, G. A. Young, G. S. Nowlan, and E. P. Dobrzanski. 2015. "Hirnantian Strata Identified in Major Intracratonic Basins of Central North America: Implications for Uppermost Ordovician Stratigraphy." *Canadian Journal of Earth Sciences* 52: 68–76.

Dobbs, D. 2005. *Reef Madness: Charles Darwin, Alexander Agassiz, and the Meaning of Coral.* New York: Pantheon.

Donovan, R. N., and M. D. Stephenson. 1991. "A New Island in the Southern Oklahoma Archipelago." *Oklahoma Geological Survey Circular* 92: 118–21.

Dott, R. H., Jr. 1974. "Cambrian Tropical Storm Waves in Wisconsin." *Geology* 2, no. 4 (May 1974): 243–46.

Dott, R. H., Jr., and C. W. Byers. 2016. "Cambrian Geology of the Baraboo Hills." *Geological Society of America Field Guide* 43: 47–54.

Droser, M. L., and S. Finnegan. 2003. "The Ordovician Radiation: A Follow-Up to the Cambrian Explosion." *Integrative and Comparative Biology* 43: 178–84.

Dukes, P. 2011. *Minutes to Midnight: History and the Anthropocene Era from 1763.* London: Anthem.

Elias, R. J., and G. A. Young. 2000. "Enigmatic Fossil Encrusting an Upper Ordovician Rocky Shore on Hudson Bay, Canada, Is a Coral." *Journal of Paleontology* 74: 179–80.

Emerson, R. W. 1836. *Nature.* James Monroe & Co., Boston & Cambridge.

Eoff, J. D. 2014. "Sequence Stratigraphy of the Upper Cambrian (Furongian, Jiangshanian and Sunwaptan) Tunnel City Group, Upper Mississippi Valley: Transgressing Assumptions of Cratonic Flooding. *Sedimentary Geology* 302: 87–101.

Fischbuch, N. R. 1970. "Devonian Reef-Building Stromatoporoids from Western Canada." *Journal of Paleontology* 44: 1071–84.

Fletcher, C. J. N. 1988. "Tidal Erosion, Solution Cavities and Exhalative Mineralization Associated with the Jurassic Unconformity at Ogmore, South Glamorgan." *Proceedings of the Geologists' Association* 99: 1–14.

Gastil, R. G., R. P. Phillips, and E. C. Allison. 1973. "Reconnaissance Geologic Map of the State of Baja California." *Geological Society of America*, no. 1: 250,000.

George, D. 2001. "On the Pilgrim's Path." *Islands* 21: 82–95.

Glenister, B. F., and W. M. Furnish. 1961. "The Permian Ammonoids of Australia." *Journal of Paleontology* 5: 673–736.

Góis-Marques, C., J. Madeira, and M. Menezes de Sequeira. 2017. "Inventory and Review of the Mio-Pleistocene São Jorge Flora (Madeira Island, Portugal): Palaeoecological and Biogeographical Implications." *Journal of Systematic Palaeontology* 16, no. 2: 159–77.

Gordon, W. A. 1973. "Marine Life and Ocean Surface Currents in the Cretaceous." *Journal of Geology* 81: 269–84.

Grabau, A. W. 1940. *The Rhythm of the Ages.* Peking: Henry Vetch.

Hagadorn, J. W., R. H. Dott Jr., and D. Damrow. 2002. "Stranded on a Late Cambrian Shoreline: Medusae from Central Wisconsin." *Geology* 30: 147–50.

Hocking, R. M., H. T. Mores, and W. J. E. Van de Graaff. 1987. *Geology of the Carnarvon Basin Western Australia*, no. 133. Perth: State Printing Division.

Ibrahim, N., S. Maganuco, C. Dal Sasso, M. Fabbri, M. Auditore, G. Bindellini, D. M. Martill, et al. 2020. "Tail-Propelled Aquatic Locomotion in a Theropod Dinosaur." *Nature* 581: 67–70.

Johnson, M. E. 1988. "Why Are Ancient Rocky Shores So Uncommon?" *Journal of Geology* 96: 469–80

——. 1992. "A. W. Graabau's Embryonic Sequence Stratigraphy and Eustatic Curve." In *Eustasy: The Historical Ups and Downs of a Major Geological Concept*, ed. R. H. Dott Jr., 111. Boulder, CO: Geological Society of America.

——. 2002. *Discovering the Geology of Baja California: Six Hikes on the Southern Gulf Coast.* Tucson: University of Arizona Press.

Johnson, M. E., and B. G. Baarli. 1987. "Encrusting Corals on a Latest Ordovician to Earliest Silurian Rocky Shore, Southwest Hudson Bay, Manitoba, Canada." *Geology* 15: 15–17.

——. 2005. "Erosion and Burial of Granite Rocky Shores in the Recent and Late Pleistocene of the Seychelles Islands: Physical and Biological Perspectives." *Journal of Coastal Research* 21: 867–79.

——. 2012. "Development of Intertidal Biotas Through Phanerozoic Time." In *Earth and Life*, ed. J. A. Talen, 63–128. Dordrecht, Netherlands: Springer Science + Business Media.

——. 2013. "Geomorphology and Coastal Erosion of a Quartzite Island: Hongdo in the Yellow Sea off the SW Korean Peninsula." *Journal of Geology* 121: 503–16.

——. 2015. "Charles Darwin in the Cape Verde and Galápagos Archipelagos: The Role of Serendipity in Development of Theories on the Ups and Downs of Oceanic Islands." *Earth Sciences History* 34, no. 2: 220–42.

Johnson, M. E., B. G. Baarli, M. Cahão, C. M. da Silva, J. Ledesma-Vázquez, E. J. Mayoral, R. S. Ramalho, and A. Santos. 2012. "Rhodoliths, Uniformitarianism, and Darwin: Pleistocene and Recent Carbonate Deposits in the Cape Verde and Canary Archipelagos." *Palaeogeography, Palaeoclimatology, Palaeoecology* 329–30: 83–100.

Johnson, M. E., B. G. Baarli, M. Cachão, E. Mayoral, R. S. Ramalho, A. Santos, and C. M. da Silva. 2018. "On the Rise and Fall of Oceanic Islands: Towards a Global Theory Following from the Pioneering Studies of Charles Darwin and James Dwight Dana." *Earth-Sciences Review* 180: 17–36.

Johnson, M. E., C.M. da Silva, A. Santos, B. G. Baarli, M. Cachão, E. J. Mayoral, A. C. Rebelo, and J. Ledesma-Vázquez. 2011. "Rhodolith Transport and Immobilization on a Volcanically Active Rcky Shore: Middle Miocene at Cabeço das Laranjas on Ilhéu de Cima (Madeira Archipelago, Portugal)." *Palaeogeography, Palaeoclimatology, Palaeoecology* 300: 113–27.

Johnson, M.E., and M. L. Hayes. 1993. "Dichotomous Facies on a Late Cretaceous Rocky Island as Related to Wind and Wave Patterns (Baja California, Mexico)." *Palaios* 8: 385–95.

Johnson, M. E., and J. Ledesma-Vázquez. 2016. *Gulf of California Coastal Ecology: Insights from the Present and Patterns from the Past*. San Diego: Sunbelt.

Johnson, M. E., J. Ledesma-Vázquez, and B. G. Baarli. 2006. "Vertebrate Remains on Ancient Rocky Shores: A Review with Report on Hadrosaur Bones from the Upper Cretaceous of Baja California (Mexico)." *Journal of Coastal Research* 22: 574–80.

Johnson, M. E., J. Ledesma-Vázquez, H. C. Clark, and J. A. Zwiebel. 1996. "Coastal Evolution of Late Cretaceous and Pleistocene Rocky Shores: Pacific Rim of Northern Baja California, Mexico." *Geological Society of America Bulletin* 108: 708–21.

Johnson, M. E., and W. S. McKerrow. 1995. "The Sutton Stone: An Early Jurassic Rocky Shore Deposit in South Wales." *Palaeontology* 38: 529–41.

Johnson, M. E. and J.-Y. Rong. 1989. "Middle to Late Ordovician Rocky Bottoms and Rocky Shores from the Manitoulin Island Area (Ontario). *Canadian Journal Earth Sci*ences 26: 642–53.

Johnson, M. E., J.-Y. Rong, and W.-B. Su. 2004. "Paleogeographic Orientation of the Sino-Korean Plate Based on Evidence for a Prevailing Silurian Wind Field." *Journal of Geology* 112: 671–84.

Johnson, M. E., J. Y. Rong, C. Y. Wang, and P. Wang. 2001. "Continental Island from the Upper Silurian (Ludfordian Stage) of Inner Mongolia: Implications for Eustasy and Paleogeography." *Geology* 29: 955–58.

Johnson, M. E., D. F. Skinner, and K. G. MacLeod. 1988. "Ecological Zonation During the Carbonate Transgression of a Late Ordovician Rocky Shore (North-Eastern Manitoba, Hudson Bay, Canada)." *Palaeogeography, Palaeoclimatology, Palaeoecology* 65: 93–114.

Johnson, M. E., A. Uchman, P. J. M. Costa, R. S. Ramalho, and S. P. Ávila. 2017. "Intense Hurricane Transports Sand Onshore: Example from the Pliocene Malbusca Section on Santa Maria Island (Azores, Portugal)." *Marine Geology* 385: 244–94.

Johnson, M. E., and G. E. Webb. 2007. "Outer Rocky Shores of the Mowanbini Archipelago, Devonian Reef Complex, Canning Basin, Western Australia." *Journal of Geology* 115: 583–600.

Johnson, M. E., G. E. Webb, B. G. Baarli, and D. R. Walsh. 2013. "Upper Devonian Shoal-Water Delta Integrated with Cyclic Back-Reef Facies off the Mowanbini Archipelago (Canning Basin), Western Australia." *Facies* 59: 991–1009.

Johnson, M. E., M. A. Wilson, and J. A. Redden. 2010. "Borings in Quartzite Surf Boulders from the Upper Cambrian Basal Deadwood Formation, Black Hills of South Dakota." *Ichnos* 17: 48–55.

Keeling, M., and S. Kershaw. 1994. "Rocky Shore Environments in the Upper Silurian of Gotland, Sweden." *GFF* 116, no. 2: 69–74.

Kenyon, W. A., ed., *The Journal of Jens Munk, 1619–1620*. Toronto: Royal Ontario Museum, 1980.

Leitão, H., and W. Alvarez. 2011. "The Portuguese and Spanish Voyages of Discovery and the Early History of Geology." *Geological Society of America Bulletin* 123: 1219–33.

Lescinsky, H. L., J. Ledesma-Vázquez, and M. E. Johnson. 1991. "Dynamics of Late Cretaceous Rocky Shores (Rosario Formation) from Baja California Mexico." *Palaios* 6: 126–41.

Lochman-Balk, C. 1964. "Paleo-Ecological Studies of the Deadwood Formation (Cambrian-Ordovician)." In *Paleontology and Stratigraphy*, ed. R. K. Sundaram, 25–38. Proceedings of Section 8 of the 22nd International Geological Congress, India.

——. 1970. "Upper Cambrian Faunal Patterns on the Craton." *Geological Society of America Bulletin* 81: 3197–3224.

MacArthur, R. H., and E. O. Wilson. 1967. *The Theory of Island Biogeography*. Princeton, NJ: Princeton University Press.

Marsaglia, K. M., and G. D. Klein. 1983. "The Paleogeography of Paleozoic and Mesozoic Storm Depositional Systems." *Journal of Geology* 91: 117–42.

Marshak, S., M. S Wilkerson, and J. DeFrates. 2016. Structural geology of the Baraboo District, 13–36, In *Geology of the Baraboo, Wisconsin Area*, ed. R. A. Davis Jr., R. H. Dott Jr. and I. W. D. Dalziel. *Geological Society of America Field Guide 43*.

Mayoral, E., A. Santos, J. A. Gámez-Vintaned, J. Ledesma-Vázquez, B. G. Baarli, M. Cachão, C. M. da Silva, and M. E. Johnson. 2018. "Upper Pleistocene Trace Fossils from Ponta das Bicudas, Santiago, Cape Verde Islands: Systematics, Taphonomy and Palaeoenvironmental Evolution." *Palaeogeography, Palaeoclimatology, Palaeoecology* 498: 83–98.

McLynn, F. 2015. *Genghis Khan: His Conquests, His Empire, His Legacy*. Boston: Da Capo.

Meireles, R. P., R. Quartau, R.S. Ramalho, A. C. Rebelo, J. Madeira, V. Zanon, and S. P. Ávila. 2013. "Depositional Processes on Oceanic Island Shelves—Evidence from Storm-Generated Neogene Deposits from the Mid-North Atlantic." *Sedimentology* 60: 1769–85.

Morris, W. J. 1981. "A New Species of Hadrosaurian Dinosaur from the Upper Cretaceous of Baja California: *Lambeosaurus laticaudus*." *Journal of Paleontology* 55: 453–62.

Muhs, D. R., and L. T. Groves. 2018. "Little Islands Recording Global Events: Late Quaternary Sea Level History and Paleozoogeogaphy of Santa Barbara and Anacapa Islands, Chanel Islands National Park, California." *Western North American Naturalist* 74: 540–89.

Neihardt, J. G. 1979. *Black Elk Speaks*. Lincoln: University of Nebraska Press.

Nelson, S. J., and M. E. Johnson. 2002. "Jens Munk Archipelago: Ordovician-Silurian Islands in the Churchill Area of the Hudson Bay Lowlands, Northern Manitoba." *Journal of Geology* 110: 577–89.

Nicoloff, P. L. 2008. *Sacred Koyasan*. Albany: State University of New York.

Payne, J. L., and M. E. Clapham. 2012. "End-Permian Mass Extinction in the Oceans: An Ancient Analogy for the Twenty-First Century?" *Annual Review of Earth and Planetary Sciences* 40: 89–111.

Percival, I. G., and B. D. Webby. 1996. "Island Benthic Assemblages: With Examples from the Late Ordovician of Eastern Australia." *Historical Biology* 11: 171–85.

Playford, P. E. 2002. "Palaeokarst, Pseudokarst, and Sequence Stratigraphy in Devonian Reef Complexes of the Canning Basin, Western Australia." In *The Sedimentary Basins of Western Australia 3*, ed. M. Keep and S. J. Moss, 763–93. Perth, Western Australia: Petroleum Exploration Society of Australia.

Playford, P. E., R. M. Hocking, and A. E. Cockbain. 2009. "Devonian Reef Complexes of the Canning Basin, Western Australia." *Geological Survey of Western Australia Bulletin* 145.

Ramalho, R. S., A. B. da Silveira, P. E. Fonseca, J. Madeira, M. Cosca, M. Cachão, M. M. Fonseca, and S. N. Prada. 2015. "The Emergence of Volcanic Oceanic Islands on a Slow-Moving Plate: The Example of Madeira Island, NE. Atlantic." *Geochemistry Geophysics Geosystems* 16, no. 2: 522–37.

Ramalho, R. S., G. Helffrich, J. Madeira, M. Cosca, C. Thoas, R. Quartau, A. Hipólito, A. Rovere, P. J. Hearty, and S. P. Ávila. 2017. "Emergence and Evolution of Santa Maria Island (Azores)—The Conundrum of Uplifted Islands Revisited." *Geological Society of America Bulletin* 129: 372–91.

Ramalho, R. S., R. Quartau, A. S. Trenhaile, N. C. Mitchell, C. D. Woodroofe, and S. P. Ávila. 2013. "Coastal Evolution on Volcanic Oceanic Islands: A Complex Interplay Between Volcanism, Erosion, Sedimentation, Sea Level Change and Biogenic Production." *Earth-Science Reviews* 127: 140–70.

Rasmussen, C. M. Ø., B. Kröger, M. L. Nielsen, and J. Colmenar. 2019. "Cascading Trend of Early Paleozoic Marine Radiations Paused by Late Ordovician Extinctions." *Proceedings of the National Academy of Sciences* 116, no. 15: 7207–13.

Raventon, E. 1994. *Island in the Plains, A Black Hills Natural History*. Boulder, CO: Johnson.

Rebelo, A. C., M. W. Rasser, A. Kroh, M. E. Johnson, R. S. Ramalho, C. Melo, A. Uchman et al. 2016. "Rocking Around a Volcanic Island Shelf: Pliocene Rhodolith Beds from Malbusca, Santa Maria Island (Azores, NE Atlantic). *Facies* 62 :1–31.

Rickards, R. B. 2000. "The Age of the Earliest Club Mosses: The Silurian Baragwanathia Flora in Victoria, Australia." *Geological Magazine* 137: 207–209.

Riosmena-Rodríguez, R., W. Nelson, and J. Aguirre, eds. 2017. *Rhodolith/Maërl Beds: A Global Perspective*. Cham, Switzerland: Springer.

Riosmena-Rodrígues, R., D.L. Steller, G. Hinojosa-Arango, and M.S. Foster. 2010. "Reefs that Rock and Roll: Biology and Conservation of Rhodolith Beds in the Gulf of California." In *The Gulf of California—Biodiversity and Conservation*, ed. R. Brusca, 49–71. Arizona-Sonora Desert Museum Studies in Natural History, University of Arizona Press and Arizona-Sonora Desert Museum, Tucson, Arizona.

Rong, J., M. E. Johnson, B. G. Baarli, W. Li, W. Su, and J. Wang. 2001. "Continental Island from the Upper Silurian (Ludlow) Sino-Korean Plate. *Chinese Science Bulletin* 46: 238–41.

Rong, J., M. E. Johnson, Z. Deng, D. Dong, S. Yaosong, B. G. Baarli, and G. Wang. 2013. "Coral-Stromatoporoid Faunas from the Shores of a Late Silurian Island, Inner Mongolia, North China." *Memoirs of the Association of Australasian Palaeontologists* 44: 95–105.

Rudkin, D. M., G. A. Young, R. J. Elias, and E. P. Dobrzanski. 2003. "The World's Biggest Trilobite—*Isotelus rex* New Species from the Upper Ordovician of Northern Manitoba, Canada." *Journal of Paleontology* 77, no. 1: 99–112

Rudkin, D. M., G. A. Young, and G. S. Nowlan. 2008. "The Oldest Horseshoe Crab: A New Xiphosurid from Late Ordovician Konservat-Lagerstätten Deposits, Manitoba, Canada." *Palaeontology* 51: 1–9.

Runnegar, B. 1979. "Ecology of *Eurydesma* and the *Eurydesma* Fauna, Permian of Eastern Australia." *Alcheringa* 3: 261–85.

Runnegar, B., J. Pojeta, M. E. Taylor, and D. Collins. 1979. "New Species of the Cambrian and Ordovician Chitons *Matthevia* and *Cheloides* from Wisconsin and Queensland: Evidence for the Early History of Polyplacophoran Mollusks." *Journal of Paleontology* 53: 1374–94.

Santos, A., E. Mayoral, B. G. Baarli, C. M. da Silva, M. Cachão, and M. E. Johnson. 2012b. "Symbiotic Association of a Pyrgomatid Barnacle and a Coral from a Volcanic Middle Miocene Shoreline (Porto Santo, Madeira Archipelago, Portugal)." *Palaeontology* 55: 173–82.

Santos, A., E. Mayoral, C. M. da Silva, M. Cahão, M. E. Johnson, and B. G. Baarli. 2011. "Miocene Intertidal Zonation on a Volcanically Active Shoreline: Porto Santo in the Madeira Archipelago, Portugal." *Lethaia* 44: 26–32.

Santos, A., E. Mayoral, C. P. Dumont, C. M. da Silva, S. P. Ávila, B. G. Baarli, M. Cachão, M. E. Johnson, R. S. Ramalho. 2015. "Role of Environmental Change in Rock-Boring Echinoid Trace Fossils." *Palaeogeography, Palaeoclimatology, Palaeoecology* 432: 1–14.

Santos, A. G., E. Mayoral, M. E. Johnson, B. G. Baarli, C. M. da Silva, M. Cachão, and J. Ledesma-Vázquez. 2012a. "Basalt Mounds and Adjacent Depressions Attract Contrasting Biofacies on a Volcanically Active Middle Miocene Shoreline (Porto Santo, Madeira Archipelago, Portugal)." *Facies* 58: 573–585.

Scarth, A. 1994. *Volcanoes: An Introduction.* London: University College Press Limited.

Schoene, B., J. Guez, A. Bartolini, U. Schaltegger, and T. J. Blackburn. 2010. "Correlating the End-Triassic Mass Extinction and Flood Basalt Volcanisms at the 100-ka Level. *Geology* 38: 387–90.

Shen, J.-W., G. E. Webb, and J. S. Jell. 2008. "Platform Margins, Reef Facies, and Microbial Carbonates: A Comparison of Devonian Reef Complexes in the Canning Basin, Western Australia, and the Guilin Region, South China." *Earth-Science Reviews* 88: 33–59.

Sheppard, T. W. 2006. "Sequence Architecture of Ancient Rocky Shorelines and Their Response to Sea-Level Change: An Early Jurassic Example from South Wales, U.K." *Journal of the Geological Society, London* 163: 595–606.

Silberling, N. J. 2008. "Biogeographic Significance of the Upper Triassic Bivalve *Monotis* in Circum-Pacific Accreted Terranes. Tectonostratigraphic Terraces of the Circum-Pacific Region." *Earth Science Series*, no. 1, 1985: 63–70.

Simms, M. J., C. T. S. Little, and B. R. Rosen. 2002. "Corals Not Serpulids: Mineralized Colonial Fossils in the Lower Jurassic Marginal Facies of South Wales." *Proceedings of the Geologists' Association* 113: 31–36.

Skinner, D. F., and M.E. Johnson. 1987. "Nautiloid Debris Oriented by Long-Shore Currents Along a Late Ordovician–Early Silurian Rocky Shore." *Lethaia* 20: 152–58.

Soja, C. M. 1992. "Potential Contributions of Ancient Oceanic Islands to Evolutionary Theory." *Journal of Geology* 100: 125–34.

Sunamura, T. 1992. *Geomorphology of Rocky Coasts.* Medford, MA: John Wiley.

Surlyk, F., and W. K. Christensen. 1974. "Epifaunal Zonation on an Upper Cretaceous Rocky Coast." *Geology* 2: 529–34.

Surlyk, F., and A. M. Sørensen. 2010. An Early Campanian Rocky Shore at Ivö Klack, Southern Sweden." *Cretaceous Research* 31: 567–76.

Thompson, P. J. 2001. *Guidebook for Geological Field Trips in New England: 2001 Annual Meeting of the Geological Society of America, Boston, Massachusetts,* ed. D. P. West Jr. and R. H. Bailey. Boston: Geological Society of America.

Toral-Granda, V. 2018. "Stowaways to Paradise." *Galapagos News* (Spring–Summer 2018): 9.

Trueman, A. E. 1922. "The Liassic Rocks of Glamorgan." *Proceedings of Geologists' Association* 33: 245–84.

Tsien, H. H. 1979. "Paleoecology of Algal-Bearing Facies in the Devonian (Couvinian to Frasnian) Reef Complexes of Belgium." *Palaeogeography, Palaeoclimatology, Palaeoecology* 27: 103–27.

Uchman, A., M. E. Johnson, R. S. Ramalho, R. Quartau, B. Berning, A. Hipólito, C. S. Melos, A. C. Rebelo, R. Cordeiro, and S. P. Ávila. 2020. "Neogene Marine Sediments and Biota Encapsulated Between Lava Flows on Santa Maria Island (Azores, North-East Atlantic): An Interplay Between Sedimentary Erosional and Volcanic Processes." *Sedimentology* 67: 3595–3618.

Uchman, A., M. E. Johnson, A. C. Rebelo, C. Melo, R. Cordeiro, R. S. Ramalho, and S. P. Ávila. 2016. "Vertically-Oriented Trace Fossil *Macaronichnus segregatis* from Neogene of Santa Maria Island (Azores; NE Atlantic) Records Vertical Fluctuations of the Coastal Groundwater Mixing Zone on a Small Oceanic Island." *Geobios* 49: 229–41.

Uchman, A., P. Torres, M. E. Johnson, B. Berning, R. S. Ramalho, A. C. Rebelo, C. S. Melo, L. Baptista, P. Madeira, R. Cordeiro, and S. P. Ávila. 2018. "Feeding Traces of Recent Ray Fish and Abundant Occurrences of the Trace Fossil *Piscichnus waitemata* from the Pliocene of Santa Maria Island Azores (Northeast Atlantic)." *Palaios* 33: 361–75.

United Nations Office For Disaster Risk Reduction. "Our World at Risk: Transforming Governance for a Resilient Future." Geneva, Switzerland: UNDRR. http://www.undrr.org/GAR2022.

USGCRP. 2018. *Impacts, Risks, and Adaptation in the United States: Fourth National Climate Assessment, Volume II: Report-in Brief*, ed. D. R. Reidmiller, C. W. Avery, D. R. Easterling, K. E. Kunkel, K. L. M. Lewis, T. K. Maycock, and B. C. Steward. Washington DC: Publication of the U.S. Global Change Program.

Vaquero, J. M., et al. 2008. "A Historical Analog of the 2005 Hurricane Vince." *American Meteorological Society* 85: 191–201.

von Humboldt, A. 1814. *Personal Narrative of Travels to the Equinoctial Regions of the New Continent during the years 1799–1804*. London: Longman, Hurst, Rees, Orme, and Brown.

Ward, P. D., J. W. Haggart, R. Mitchell, J. L. Kirschvink, and T. Tobin. 2012. "Integration of Macrofossil Biostratigraphy and Magnetostratigraphy for the Pacific Coast Upper Cretaceous (Campanian-Maastrichtian) of North America and Implications for Correlation with the Western Interior and Tethys." *Geological Society of America Bulletin* 124, nos. 5–6 (May 1, 2012): 957–74.

Weaver, A. J. 1990. "Ocean Currents and Climate." *Nature* 374: 432.

White, R. S., and J. K. Greenberg. 2016. "Religious Faith as a Motivation in Using Geosciences to Develop a Sustainable Future. In *Geoscience for the Public Good and Global Development: Toward a Sustainable Future*, ed. G. R. Wessel and J. K. Greeenberg, 23–34. Boulder, CO: Geological Society of America.

Wilkes, C., ed. *United States Exploring Expedition During the Years 1838, 1839, 1840, 1841*. New York: George Putnam, 1849.

Wilson, M.A., and P. D. Taylor. 2001. "Palaeoecology of Hard Substrate Faunas from the Cretaceous Quhlah Formation of the Oman Mountains." *Palaeontology* 44: 21–41.

Young, G. A., D. M. Rudkin, E. P. Dobrzanski, S. P. Robson, and G. S. Nowlan. 2007. "Exceptionally Preserved Late Ordovician Biotas from Manitoba, Canada." *Geology* 35, no. 10: 883–86.

INDEX

Page numbers in *italics* indicate photos. Page numbers in **bold** indicate glossary terms.